MW00845079

ELECTRICITY 4

AC/DC MOTORS, CONTROLS, AND MAINTENANCE

10TH EDITION

ELECTRICITY 4

AC/DC MOTORS, CONTROLS, AND MAINTENANCE

10TH EDITION

JEFF KELJIK

DELMAR
CENGAGE Learning

Australia • Brazil • Japan • Korea • Mexico • Singapore • Spain • United Kingdom • United States

Electricity 4: AC/DC Motors, Controls, and Maintenance, 10E
Jeff Keljik

Vice President, Editorial: Dave Garza

Director of Learning Solutions: Sandy Clark

Acquisitions Editor: Jim DeVoe

Managing Editor: Larry Main

Senior Product Manager: John Fisher

Editorial Assistant: Aviva Ariel

Vice President, Marketing: Jennifer Baker

Marketing Director: Deborah Yarnell

Marketing Development Manager: Erin Brennan

Marketing Brand Manager: Kristin McNary

Senior Production Director: Wendy Troeger

Production Manager: Mark Bernard

Content Project Manager: Barbara LeFleur

Senior Art Director: David Arsenault

For product information and technology assistance, contact us at
Cengage Learning Customer & Sales Support, 1-800-354-9706

For permission to use material from this text or product, submit all requests online at **www.cengage.com/permissions**. Further permissions questions can be e-mailed to **permissionrequest@cengage.com**

Microsoft® is a registered trademark of the Microsoft Corporation.

Library of Congress Control Number: 2012944012

ISBN-13: 978-1-111-64675-2

ISBN-10: 1-111-64675-9

Delmar
5 Maxwell Drive
Clifton Park, NY 12065-2919
USA

Cengage Learning is a leading provider of customized learning solutions with office locations around the globe, including Singapore, the United Kingdom, Australia, Mexico, Brazil, and Japan. Locate your local office at: **international.cengage.com/region**

Cengage Learning products are represented in Canada by Nelson Education, Ltd.

To learn more about Delmar, visit **www.cengage.com/delmar**
Purchase any of our products at your local college store or at our preferred online store **www.cengagebrain.com**

Notice to the Reader
Publisher does not warrant or guarantee any of the products described herein or perform any independent analysis in connection with any of the product information contained herein. Publisher does not assume, and expressly disclaims, any obligation to obtain and include information other than that provided to it by the manufacturer. The reader is expressly warned to consider and adopt all safety precautions that might be indicated by the activities described herein and to avoid all potential hazards. By following the instructions contained herein, the reader willingly assumes all risks in connection with such instructions. The publisher makes no representations or warranties of any kind, including but not limited to, the warranties of fitness for particular purpose or merchantability, nor are any such representations implied with respect to the material set forth herein, and the publisher takes no responsibility with respect to such material. The publisher shall not be liable for any special, consequential, or exemplary damages resulting, in whole or part, from the readers' use of, or reliance upon, this material.

Printed in the United States of America
3 4 5 6 7 8 9 22 21 20 19 18

CONTENTS

PREFACE

The tenth edition of *Electricity 4* has been updated to provide more information and better flow of concepts. New material and artwork have been added to better reflect the current workplace. At the same time, the text has retained the features and style of previous editions that have made it so popular.

The text introduces the concepts of AC and DC motors, and the associated controls and maintenance of this equipment. The material is broken down into short segments that concentrate on specific concepts or application of particular types of equipment. The detailed explanations are written in easy-to-understand language that concisely presents the required knowledge. Many illustrations and photographs help provide technical understanding and provide real-world references. This type of explanation and application better prepares the student to perform effectively on the job in installing, troubleshooting, repairing, and servicing electrical motors and controls.

The knowledge obtained in this book permits the student to progress further in the study of electrical systems. The study of electricity and the application of electrical products are continually changing. The electrical industry constantly introduces new and improved devices and material, which in turn lead to changes in installing and operating equipment. Electrical codes also change to reflect the industry needs. It is essential that students continue to learn and update their knowledge of current procedures and practices.

The text is easy to read and the units have been grouped by general topics. Summaries of each unit provide an opportunity to restate the most important topics of the unit. Summary Reviews of the units reemphasize topic groups.

Each unit begins with learning objectives. An Achievement Review at the end of each unit provides an opportunity for readers to check their understanding of the material in small increments before proceeding. Some of the problems in the text require the use of simple algebra, and the student should be familiar with the math before trying to solve the equations. It is also essential that the reader have a basic understanding of the fundamentals of electrical circuits and electrical concepts.

It is recommended that the most recent edition of the *National Electrical Code*® (published by the National Fire Protection Association) be available for reference and use as the learner applies this text. Applicable state and local codes and regulations should also be consulted when making the actual installations.

Features of the tenth edition include

- updated coverage on DC motor starting.
- expanded coverage of Servo motors.
- updated information on electronic controls.
- an update to the 2011 *NEC*®.

An instructor's guide for *Electricity 4* is available. The guide includes the answers to the Achievement Reviews and Summary Reviews and additional test questions. Instructors may use these questions to devise additional tests to evaluate student learning.

INSTRUCTOR SITE

An Instruction Companion Website containing supplementary material is available. This site contains an Instructor Guide, testbank, image gallery of text figures, and chapter presentations done in PowerPoint. Contact Delmar Cengage Learning or your local sales representative to obtain an instructor account.

Accessing an Instructor Companion Website site from SSO Front Door

1. Go to http://login.cengage.com and log in using the Instructor email address and password.
2. Enter author, title, or ISBN in **the Add a title to your bookshelf** search box, and click on the **Search** button.
3. Click **Add to My Bookshelf** to add Instructor Resources.
4. At the Product page, click on the **Instructor Companion site** link.

New Users

If you're new to Cengage.com and do not have a password, contact your sales representative.

ABOUT THE AUTHOR

Jeff Keljik has been teaching at Dunwoody Institute in Minneapolis for more than 33 years, where he was the head of electrical programs for more than 16 years. He is a licensed master and journeyman electrician with a bachelor's degree in business communication. He currently teaches classes for corporate clients locally and nationally. Jeff teaches classes for journeyman and master electricians to update their licenses as well as teaching apprenticeship classes on-line. He also manages the electrical construction and maintenance projects for the college campus buildings at Dunwoody Institute. He has worked as an electrician and as a consultant on international training projects in the electrical industry.

In addition to his teaching and administrative positions, Mr. Keljik serves the North Central Electrical League as Chairman of the Board of Directors and on the Education committee. He also serves as an advisor on the education committee for the Minnesota Electrical Association (MEA). He has written several texts on motor and motor control systems, power generation and distribution, and on fundamental DC and AC theory. His books include: *Electricity 3* and *Electric Motors and Motor Controls.* Jeff has co-written books on basic electrical concepts in DC and AC electrical theory.

ACKNOWLEDGMENTS

The authors and Delmar Cengage Learning would like to take this opportunity to acknowledge those who contributed to the review process for this edition of *Electricity 4:*

David Adams
Niagara County Community College
Sanborn, NY

Joseph T. Brown
Tri County Technical College
Pendleton, SC

Phillip Serina
Kaplan Career Institute
Brooklyn, Ohio

DEDICATION

I would like to dedicate this tenth edition to my family and all of my friends. My friends in the electrical industry help with ideas, and friends outside the industry provide support and enthusiasm.
                                                                                    —*Jeff Keljik*

ELECTRICAL TRADES

The Delmar series of instructional material for the electrical trades includes the texts, text workbooks, and related information workbooks listed below. Each text features basic theory with practical applications and student involvement in hands-on activities.

Electricity 1

Electricity 2

Electricity 3

Electricity 4

Electric Motor Control

Electric Motor Control
   Laboratory Manual

Industrial Motor Current

Alternating Current
   Fundamentals

Electrical Wiring—
   Residential

Electrical Wiring—
   Commercial

Electrical Wiring—
   Industrial

Practical Problems
   In Mathematics
   For Electricians

Equations based on Ohm's law.

P = Power in watts

I = Intensity of current
   in amperes

R = Resistance in ohms

E = Electromotive force
   in volts

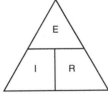

$E = IR$

$I = \dfrac{E}{R}$

$R = \dfrac{E}{I}$

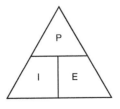

$P = IE$

$I = \dfrac{P}{E}$

$E = \dfrac{P}{I}$

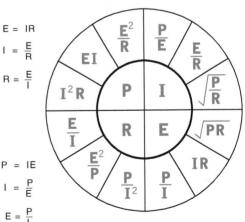

© Cengage Learning 2014

UNIT

THE DC SHUNT MOTOR

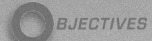

OBJECTIVES

After studying this unit, the student should be able to

- list the parts of a DC shunt motor.
- draw the connection diagrams for series shunt and compound motors.
- define torque and tell what factors affect the torque of a DC shunt motor.
- describe counter-EMF (CEMF) and its effects on current input.
- describe the effects of an increased load on armature current, torque, and speed of a DC shunt motor.
- list the speed control, torque, and speed regulation characteristics of a DC shunt motor.
- make DC motor connections.

1

The production of electrical energy and its conversion to mechanical energy in electric motors of all types is the basis of industrial productivity. Very basic DC motor principles are given in *Electricity 1*.

CONSTRUCTION FEATURES

Conventional DC motors closely resemble DC generators in construction features. In fact, it is difficult to identify them by appearance alone. A motor has the same two main parts as a generator—the field structure and the armature assembly consisting of the armature core, armature winding, commutator, and brushes. Some general features of a DC motor are shown in Figure 1–1.

CARBON BRUSH

FIGURE 1–1 (A) DC motor armature with commutator bars. (B) Permanent-magnet DC motor with rotor and carbon brush connections.

The Field Structure

The field structure of a motor has at least two pairs of field poles, although motors with four pairs of field poles are also used, as shown in Figure 1–2(A). A strong magnetic field is provided by the field windings of the individual field poles. The magnetic polarity of the field system is arranged so that the polarity of any particular field pole is opposite to that of the poles adjacent to it.

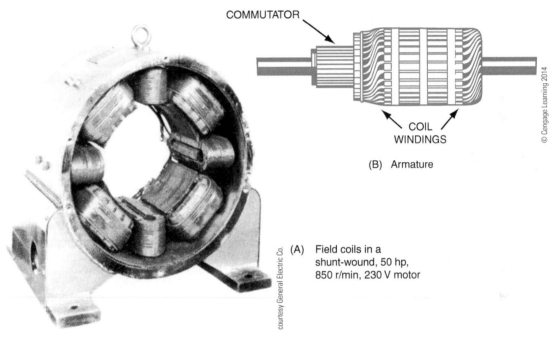

COMMUTATOR

COIL WINDINGS

(B) Armature

(A) Field coils in a shunt-wound, 50 hp, 850 r/min, 230 V motor

courtesy General Electric Co.

© Cengage Learning 2014

FIGURE 1–2 Field structure and armature assembly of a motor.

The Armature

The armature of a motor is a cylindrical iron structure mounted directly on the motor shaft, as shown in Figure 1–2(B). In DC motors, the armature is the rotating component of the motor. Armature windings are embedded in slots in the surface of the armature and terminate in segments of the commutator. Current is fed to these windings on the rotating armature by carbon brushes that press against the commutator segments. This current in the armature windings sets up a magnetic field in the armature that reacts with the magnetic field of the field poles. These magnetic effects are used to develop torque, which causes the armature to turn (Figure 1–3). The commutator changes the direction of the current in the armature conductors as they pass across poles of opposite magnetic polarity. Continuous rotation in one direction results from these reversals in the armature current.

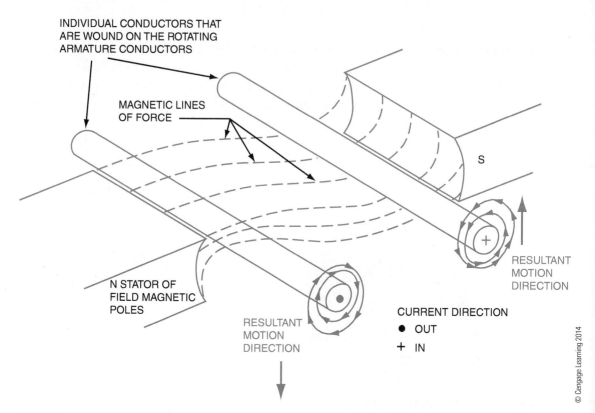

INDIVIDUAL CONDUCTORS THAT
ARE WOUND ON THE ROTATING
ARMATURE CONDUCTORS

MAGNETIC LINES
OF FORCE

S

N STATOR OF
FIELD MAGNETIC
POLES

RESULTANT
MOTION
DIRECTION

RESULTANT
MOTION
DIRECTION

CURRENT DIRECTION
● OUT
+ IN

© Cengage Learning 2014

FIGURE 1–3 Torque, or force direction, on a current-carrying conductor in a magnetic field.

Figure 1–4 shows a cutaway view of a DC motor available with horsepower ratings ranging from 25.0 hp to 1000 hp.

By using the right-hand rule for motors, as illustrated in Figure 1–5, you can determine which direction a current-carrying conductor will move when placed in a magnetic field. This is the principle of all motor action. The first finger represents the direction of the stator flux, as magnetic lines travel north to south. The center finger represents the direction of the electron current in the conductor placed within the magnetic field. The thumb represents the thrust of the conductor as it tries to move out of the magnetic field. As shown in Figure 1–3, the coils of the armature are spaced around the rotor, and the rotor conductors are connected to the commutator segments. Because the coils of wire on the armature (rotor) have current flow through them, the magnetic field thus created in the rotor conductors reacts with the stationary field (stator), and the conductors will move according to the right-hand rule for motors. The conductors of the rotor that are directly adjacent to the stator pole pieces have the maximum rotor current and therefore the most magnetic interaction with the main magnetic poles.

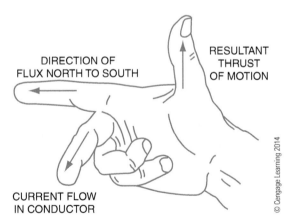

1. Main shaft
2. Bearings
3. Grease "meter"
4. Ventilating fan
5. Armature banding
6. Armature equalizer coil assembly
7. Lifting lugs
8. Frame
9. Inspection plate
10. Main field coil
11. Commutating coils
12. Main field coil
13. Armature
14. Commutator connections to
    armature turns
15. Commutator
16. Brushholder
17. Brushholder yoke
18. Mounting feet
19. Terminal conduit box

Courtesy of General Electric, DC Motor and Generator Department

FIGURE 1-4 Assembled 23 hp DC motor.

DIRECTION OF
FLUX NORTH TO SOUTH

RESULTANT
THRUST
OF MOTION

CURRENT FLOW
IN CONDUCTOR

© Cengage Learning 2014

FIGURE 1-5 Right-hand rule for motors using electron flow.

TYPES OF DC MOTORS

Shunt, series, compound, and permanent magnet motors are all widely used. The schematic diagrams for each type of motor are shown in Figure 1–6. The selection of the type of motor to use is based on the mechanical requirements of the applied load. A *shunt motor* has the field

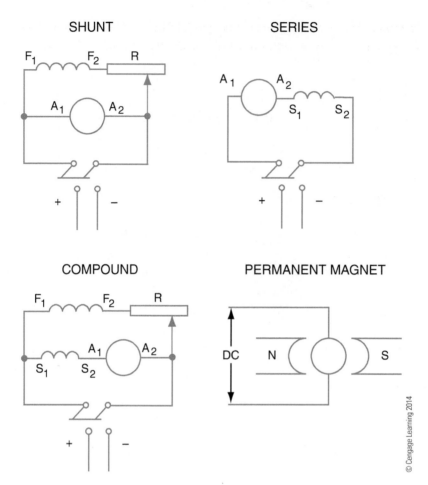

FIGURE 1-6 Schematic diagrams that show motor connections.

circuit connected in shunt (parallel) with the armature, whereas a *series motor* has the armature and field circuits in series. A *compound motor* has both a shunt and a series field winding. A *permanent magnet motor* only has armature connections.

MOTOR RATINGS

DC motors are rated by their voltage, current, speed, and horsepower output. The number and methods of connection for the armature and field also dictate motor operating characteristics.

TORQUE

The rotating force at the motor shaft produced by the interaction of the magnetic fields of the armature and the field poles is called *torque*. As the magnitude of the torque increases, the twisting force of the shaft increases. Torque is defined as the product of the force in pounds and the radius of the shaft or pulley in feet.

For example, a motor that produces a tangential force of 120 pounds at the surface of the shaft 2 in. in diameter or 1 in. in radius has a torque of 10 foot-pounds (ft-lbs).

$$\text{Torque} = \text{Force} \times \text{Radius}$$

$$= 120 \text{ lbs} \times 1/12 \text{ ft} = 10 \text{ ft-lbs}$$

Torque in a motor depends on the magnetic strengths of the field and the armature. The torque increases along with the armature current and, consequently, the strength of the armature magnetic field increases.

Another example is shown in Figure 1–7. If a V-belt drive has a pulling force at the radius of a motor pulley of 25 pounds, and the pulley surface is 1 ft from the center of the motor shaft, the motor torque is the product of the radius (measured in ft) and the pulling force (measured in pounds). The torque is usually measured in foot-pounds, the standard measure for torque. For small motors, the torque may be measured in ounce-inches (oz-in). The same principle is used, but the ounce-inch measurement must be divided by 192 (12 inches × 16 ounces) to get the equivalent foot-pounds.

It is necessary to distinguish between the torque developed by a motor when operating at its rated speed and the torque developed at the instant the motor starts. Certain types of motors have high torque at rated speed but poor starting torque. The many types of loads that can be applied to motors mean that the torque characteristic must be considered when selecting a motor for a particular installation.

ROTATION

The direction of the armature rotation of a DC motor depends on the direction of the current in the field circuit and the armature circuit (Figure 1–8). To reverse the direction of rotation, the current direction in *either* the field or the armature must be reversed. Reversing the power supply leads causes both the armature and the field to become reversed, as shown in Figure 1–9. To determine the direction of the conductor movement, use the right-hand rule for motors, as shown in Figure 1–5. The first finger indicates the direction of the flux (north to south), the center finger indicates the direction of the current flow (negative to positive), and the thumb indicates the direction of the resultant thrust.

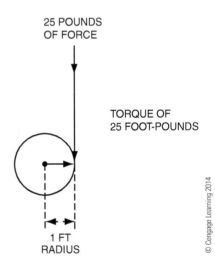

25 POUNDS OF FORCE

TORQUE OF 25 FOOT-POUNDS

1 FT RADIUS

© Cengage Learning 2014

FIGURE 1–7 Torque is measured as the pulling force at some distance from the center of the motor shaft. If the force is 25 pounds and the distance is 1 ft from the center of the shaft, the motor torque is 25 ft-lbs.

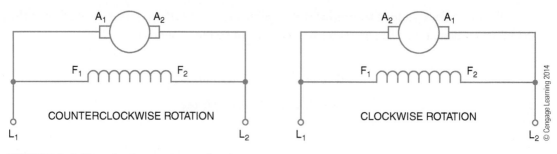

FIGURE 1-8 Standard connections for shunt motors.

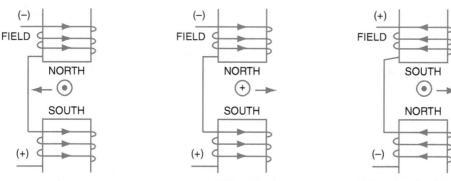

ORIGINAL CONNECTIONS GIVE FIELD POLARITY AS SHOWN. ARMATURE CURRENT AS SHOWN WOULD PRODUCE A COUNTERCLOCKWISE ROTATION.

ARMATURE CONNECTION CHANGED TO GIVE OPPOSITE DIRECTION OF CURRENT IN ARMATURE WHILE MAINTAINING FIELD DIRECTION RESULTS IN CLOCKWISE ROTATION.

REVERSE FIELD POLARITY AND ARMATURE POLARITY FROM MIDDLE DIAGRAM RESULT IN THE SAME CLOCKWISE DIRECTION OF ROTATION.

FIGURE 1-9 Reversing either the armature connections or the field connections causes the direction of armature rotation to change; changing both connections results in the same direction of rotation.

STARTING CURRENT AND CEMF

The starting current of a DC motor is much higher than the running current while the motor is operating at its rated speed. At the instant power is applied, the armature is motionless and the armature current is limited only by the very low armature circuit wire resistance. As the motor builds up to its rated speed, the current input decreases until the motor reaches its rated speed. At this point, the armature current stops decreasing and remains constant.

Factors other than armature wire resistance also limit the current. Figure 1–10 illustrates a demonstration that shows the "generator" action (CEMF) within a motor that accounts for the decrease in current with a speed increase.

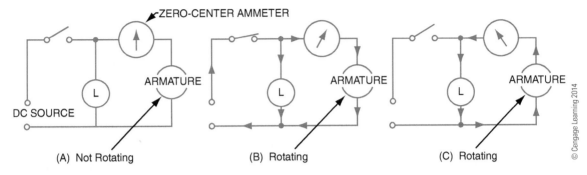

FIGURE 1–10 Demonstration of CEMF.

In Figure 1–10, a DC motor and a lamp (each with the same voltage rating) are connected in parallel to the DC source. A zero-center ammeter connected in the circuit indicates the amount and direction of the current to the motor. When the line switch is open (A), there is no current in any part of the circuit. When the switch is closed (B), the lamp lights instantly and the ammeter registers high current to the motor. The motor current decreases as the motor speed increases and remains constant when the motor reaches its rated speed. The instant the switch is opened, the ammeter deflection reverses. The lamp continues to light but grows dimmer as the motor speed falls.

Two conclusions can be drawn from this demonstration:

- A DC motor develops an induced voltage while rotating.
- The direction of the induced voltage is opposite to that of the applied voltage and for this reason is called counter-EMF (CEMF).

The amount of voltage generated within a spinning armature depends on the speed of the rotation and the strength of the magnetic field. Just as in any generator, the left-hand rule for generators applies, as shown in Figure 1–11. Even though the armature is being caused to spin through motor action, the act of spinning a coil of wire within a magnetic field causes it to act like a generator. The thumb represents the direction of thrust of the moving conductor, and the first finger represents the direction of the main stator field magnetic flux. Now the first finger represents the direction of the induced current flow within the armature conductors, which is counter to the applied current. This counter-induced potential is CEMF. The induced current flow is smaller in

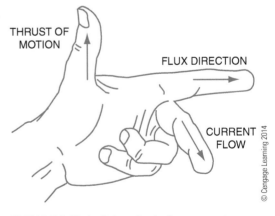

FIGURE 1–11 Left-hand rule for generators.

magnitude than the applied current flow from the voltage source. The difference between the applied current flow and the counter-induced current is the differential current.

As the torque, or twisting effort, rotates the armature, the conductors of the armature cut the main field magnetic flux, as in a generator. This action induces a voltage into the armature windings that opposes line voltage.

The production of CEMF in a DC motor accounts for the changes in current to a motor armature at different speeds. When there is no current in the circuit, the motor armature is motionless and the CEMF is zero. The starting current is very high because only the ohmic resistance of the armature limits the current. As the armature starts to rotate, the CEMF increases and the line current decreases. When the speed stops increasing, the value of the CEMF approaches the value of the applied voltage but is never equal to it. The value of the voltage that actually forces current through the motor is equal to the difference between the applied voltage and the CEMF. At rated speed, this voltage differential just maintains the motor at constant speed, as shown in Figure 1–12.

When a mechanical load is then applied to the motor shaft, both the speed and CEMF decrease. However, the voltage differential *increases* and causes an increase of input current to the motor. Any further increase in mechanical load produces a proportional increase in input current (Figure 1–13).

The increase in motor current due to an increase in mechanical load also can be justified in terms of the torque. Because torque depends on the strength of the magnetic field of the armature, which, in turn, depends on the armature current, any increase in mechanical load requires an increase in the armature current, more load, slower speed, higher differential, and more current.

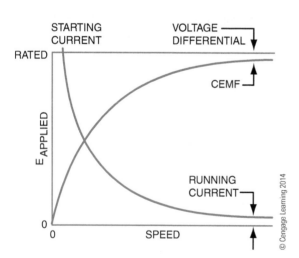

FIGURE 1–12 Effects of CEMF on the armature current.

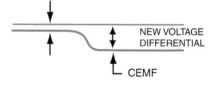
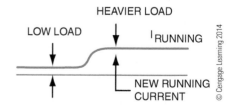

FIGURE 1–13 Effects of CEMF and $I_{armature}$ when the load is increased.

Because the starting current may be many times greater than the rated current under full load, large DC motors must not be connected directly to the power line for start-up. The heavy current surges produce excessive line voltage drops that may damage the motor. The maximum branch-circuit fuse size for any DC motor is based on the full-load running current of the motor. Therefore, starters for DC motors generally limit the starting current to 150% of the full-load running current.

ARMATURE REACTION

Armature reaction occurs in DC motors and is caused by the stator magnetic field being distorted, or altered, in reaction to the armature magnetic field. The armature reaction is actually a bending of the motor magnetic field so that the brushes are no longer aligned with the neutral magnetic plane of the motor. If the brushes are not in alignment with this magnetic plane, the current conducted to the armature does not split equally in the armature conductors and therefore causes a voltage difference at the brushes. This causes sparking where the brush meets the commutator. In a motor with a constant load, the brushes can be shifted back into the neutral plane to reduce sparking. The brushes are shifted in the direction opposite to the rotation. If the motor has a varying load, the neutral plane will be constantly shifting. To counteract the effects of the field distortion, some motors are designed with *interpoles* or *commutating poles*. These poles are connected in series with the armature circuit. Every change in armature current that would tend to distort the magnetic field is counteracted by the interpole magnetic field or commutating windings (refer to Figure 1–2).

SPEED CONTROL AND SPEED REGULATION

The terms *speed control* and *speed regulation* should not be used interchangeably. The meaning of each is entirely different. *Speed regulation* refers to a motor's capability to maintain a certain speed under varying mechanical loads from no load to full load. It is expressed as a percent. The formula used is

$$\text{Percent speed regulation} = \frac{\text{No-load speed} - \text{Full-load speed}}{\text{Full-load speed}} \times 100$$

Using this formula, you can determine that a motor that holds a constant speed between no load and full load has a 0% speed regulation.

Speed control refers to changing the motor speed intentionally by means of external control devices. This is done in a variety of ways and is not a result of the design of the motor.

Speed Control

DC motors are operated below normal speed by reducing the voltage applied to the armature circuit. Resistors connected in series with the armature may be used for voltage reduction, or electronic speed control is used to reduce the voltage applied to the armature circuit. When the armature voltage is reduced while keeping the field current constant, the CEMF is too high, and no current flows to the rotor. The rotor torque is reduced, and the speed slows (Figure 1–13). The speed of a DC motor can also be brought below its rated speed by varying the voltage applied to the entire motor. However, this method is not used because there is a loss of torque along with the reduction in speed.

A DC motor may be operated above its rated speed by reducing the strength of the stator field flux. A system is used to reduce the field current and, in turn, the field flux.

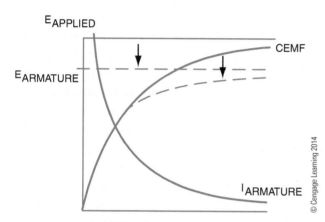

FIGURE 1–14 To reduce speed, reduce the armature voltage while keeping the field current constant.

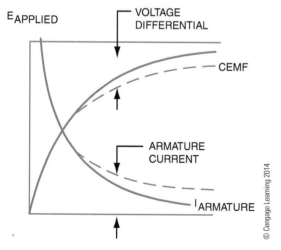

FIGURE 1–15 To increase speed, reduce the field current while keeping the armature voltage constant.

Although it seems reasonable that a reduction in field flux reduces the speed, the speed actually increases because the reduction of flux reduces the CEMF and permits the applied voltage to increase the armature current. The speed continues to increase until the increased torque is balanced by the opposing torque of the mechanical load. When the field flux is reduced while keeping the armature voltage constant, the CEMF in the armature drops (Figure 1–14). As a result, there is a larger voltage differential, which causes an increase in armature current. This develops more torque to increase the speed of the motor (Figure 1–15).

Caution: Because motor speed increases with a decrease in field flux, the field circuit of a motor should never be opened when the motor is operating, particularly when it is running freely without a load. An open field may cause the motor to rotate at speeds that are dangerous to both the machine and to the personnel operating it. For this reason, some motors are protected against excessive speed by a "no field release" feature. This device disconnects the motor from the power source if the field circuit opens.

THE SHUNT MOTOR

Two factors are important when selecting a motor for a particular application: (1) the variation of the speed with a change in load and (2) the variation of the torque with a change in load. A shunt motor is basically a constant speed device. If a load is applied, the motor *tends* to slow down. The slight loss in speed reduces the CEMF and results in an increase of the armature current. This action continues until the increased current produces enough torque to meet the demands of the increased load. As a result, the shunt motor reaches a state of stable equilibrium because a change of load always produces a reaction that adapts the power input to the change in load.

The basic circuit for a shunt motor is shown in Figure 1–16(A). Note that only a shunt field winding is shown. Figure 1–16(B) shows the addition of a series winding to counteract the effects of armature reaction. From the standpoint of a schematic diagram, Figure 1–16(B) looks like a compound motor. However, this type of motor is not considered a compound motor because the commutating winding is not wound on the same pole as the field winding, and the series field has only a few turns of wire in series with the armature circuit. As a result, the operating characteristics are those of a shunt motor. This is so noted on the nameplate of the motor by the terms *compensated shunt motor* or *stabilized shunt motor.*

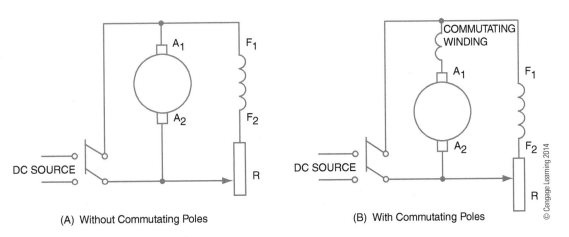

(A) Without Commutating Poles   (B) With Commutating Poles

FIGURE 1–16 Shunt motor connections.

Speed Control

A DC shunt motor has excellent speed control. To operate the motor above its rated speed, reduce the field current and field flux. To operate below rated speed, reduce the voltage applied to the armature circuit.

Electronic speed control systems are used extensively. The principles of control are the same as the old manual controls. Speeds above normal are achieved by reducing the field voltage electronically, and speeds below normal reduce the voltage ap plied to the armature. (For more detail on electronic speed control, see Unit 6 on speed control.)

Rotation

The direction of armature rotation may be changed by reversing the direction of current in either the field circuit or the armature circuit. For a motor with a simple shunt field circuit, it may be convenient to reverse the field circuit lead. If the motor has a series winding, or an interpole winding to counteract armature reaction, the same *relative* direction of current must be maintained in the shunt and series windings. For this reason, it is always safer to reverse the direction of just the armature current.

Torque

A DC shunt motor has high torque at any speed. At start-up, a DC shunt motor develops 150% of its rated torque if the starting controller is capable of withstanding the heating effects of the current. For very short periods of time, the motor can develop 350% of full-load torque, if necessary.

Speed Regulation

The speed regulation of a shunt motor drops from 5% to 10% from the no-load state to full load. As a result, a shunt motor is superior to the series DC motor, but is inferior to a compound-wound DC motor (see Units 2 and 3). Figure 1–17 shows a typical DC motor with horsepower ratings ranging from 1 hp to 5 hp.

Courtesy of General Electric, DC Motor and Generator Department

FIGURE 1-17 DC motor, 1 hp to 5 hp.

PERMANENT MAGNET MOTORS

A variation on the DC shunt motor principle is the permanent magnet (PM) motor. Two varieties are available. One style of PM motor uses a permanently magnetized material such as Alnico or ceramic magnets mounted in the stator to provide a constant magnetic field. The rotor is supplied with DC through a brush and commutator system. The result is similar to a DC shunt-type motor, but it has a very linear speed/torque curve.

Another type of PM motor uses the PMs mounted in the rotor. Because DC is still supplied to the motor, commutation must be provided to properly magnetize the stator in relation to the rotor for rotational torque. The commutator segments are actually connected to the stator windings, and a set of sliding contacts on the rotor provides the proper electrical connection from the DC source to the proper commutator segments on the stator. This type of PM motor can be produced in larger horsepower models than the PM stator types. PM motors are generally smaller than 5 hp. Figure 1–18 compares the physical size of a PM motor to a shunt motor of similar horsepower.

Courtesy of Bodine Electric Company

FIGURE 1–18 DC permanent magnet stator.

The DC permanent magnet motor is a popular motor where battery power provides the DC for the motor power. Small motors used for electrical drives in the automobile and motors such as trolling motors are good examples of PM motors. Rare-earth magnetic materials, such as samarium-cobalt and neodymium, have made the permanent magnets more powerful and allowed the size of PM motors to increase in horsepower to more than 15 hp. The torque-to-weight ratio and the horsepower-to-weight ratio make these motors a popular choice for mobile vehicles where horsepower is needed with light weight. The operating characteristics of these motors resemble the characteristics of a separately excited shunt motor. The speed curve is dependent on the armature field current.

If a motor is not providing rotational torque, the problem could be that the magnets have lost some of the original magnetic strength. Another problem that can occur is the demagnetization of the PM material. This can happen when the motors are running in one direction and then are quickly reversed under power. Some control circuits provide protection from quick reversals; others compensate for this problem by applying a small voltage during reversing.

BRUSHLESS DC MOTORS

Instead of using mechanical commutation to supply a field and power to the rotor, electronics can be used to switch the stator field. The rotor uses a PM so that no direct power is supplied to the rotor. To switch the power supply to the field windings, sensing devices must be used to determine rotor movement. As the rotor speed increases or decreases, the sensor relays the information to the electronic switching supply. The electronic supply constantly adjusts to provide the proper level of voltage to the proper stator poles to maintain speed and direction (see Figure 1–19).

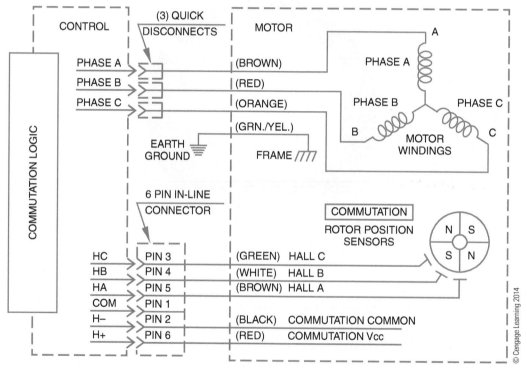

© Cengage Learning 2014

FIGURE 1–19 Brushless DC motor control schematic.

Summary

The DC shunt motor uses the shunt field as the main magnetic field in the stator. The shunt field is made up of many turns of small wire and is connected, or *shunted*, across the armature. The shunt field may have a series-connected control that adjusts the amount of current to the field. The principle of the DC motor relies on the concept of commutation. This commutator and brush connection always keep the direction of the current and the direction of the magnetic field consistent. The speed and the current to the rotor are inversely proportional. If the rotor is spinning faster, more CEMF is produced and less voltage differential, and therefore less armature current is produced. DC motors are used in a variety of styles for different purposes. Many variations of the shunt motor are used in specialized applications.

ACHIEVEMENT REVIEW

A. Select the correct answer for each of the following statements, and place the corresponding letter in the space provided.

1. DC motors are rated in _____
    a. voltage, frequency, current, and speed.
    b. voltage, current, speed, and torque.
    c. voltage, current, and horsepower.
    d. voltage, current, speed, and horsepower.

2. The generator effect in a motor produces a _____
    a. high power factor.
    b. high resistance.
    c. CEMF.
    d. reduced line voltage.

3. A DC motor draws more current with a mechanical load applied to its shaft because the

    _____
    a. CEMF is reduced with the speed.
    b. voltage differential decreases.
    c. applied voltage decreases.
    d. torque depends on the magnetic strength.

4. The direction of rotation of a compound interpole motor may be reversed by reversing the direction of current flow through the _____
    a. armature.
    b. armature or field circuit.
    c. armature, interpoles, and series field.
    d. shunt field.

5. The speed of a DC motor may be reduced below its rated without losing torque by reducing the voltage at the _____

    a. motor.
    b. series field.
    c. armature.
    d. armature and field.

6. Advantages of DC motors include _____

    a. simplicity in construction.
    b. speed control above and below base speed.
    c. excellent torque and speed control.
    d. horsepower for size.

B. Complete the following statements:

7. The twisting force exerted on the shaft of a motor is called _____ and is due to the magnetic field interaction of the _____ and _____.

8. Field interpoles connected in series with the armature circuit of a motor help counteract the effects of _____.

9. As a DC motor comes up to its rated speed, its armature current (decreases, remains the same, increases). (Circle the correct answer.)

10. The main factor controlling the armature current of a DC shunt motor operating at rated speed is the _____

    _____.

UNIT 2

THE DC SERIES MOTOR

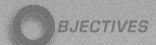

OBJECTIVES

After studying this unit, the student should be able to

- draw the basic connection circuit of a series DC motor.
- describe the effects on the torque and speed with a change in current.
- describe the effects of a reduction of a load on the speed of a DC series motor.
- connect a DC series motor.

19

Despite the wide use of alternating current for power generation and transmission, the DC series motor is often used, such as for starter motors in automobiles and aircraft. This type of motor is also used as a traction motor because of its capability to produce a high torque with only a moderate increase in power at reduced speed.

The basic circuit for the series motor is shown in Figure 2–1. The field circuit has comparatively few turns of large-diameter wire that permit it to carry the full-load current of the motor.

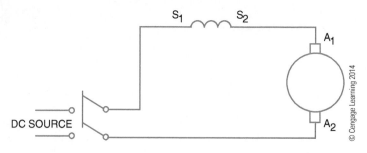

FIGURE 2–1 Series motor connections.

TORQUE

A series motor develops up to 500% of its full-load torque at starting. Therefore, this type of motor is used for railway installations, cranes, and other applications requiring high torque at low speed. The series motor is used in many electric locomotives, where the shaft becomes the axle for the drive wheels.

Remember that the shunt motor operates at constant speed. In a shunt motor, any increase in torque requires a proportionate increase in armature current. In a series motor, the stator field is operated below magnetic saturation, and any increase in load causes an increase of current in both the field and armature circuits. As a result, the armature flux and the field flux increase together. Because torque depends on the interaction of the armature and field fluxes, the torque increases as the square of the value of the current increases. Therefore, a series motor produces a greater torque than a shunt motor for the same increase in current. The series motor, however, shows a greater reduction in speed as mechanical load is added. A light load has little current draw, and the armature and field current are reduced.

SPEED CONTROL AND SPEED REGULATION

The speed regulation of a line-connected series motor is inherently poorer than that of a shunt motor. If the motor is running with a mechanical load and the load is suddenly reduced, the motor speed increases. The increase in speed creates more CEMF in the armature, and the circuit current is reduced. The circuit current is the same current in both the stator field and the rotor field, therefore squaring the effect of the current reduction. The torque that is produced drops off dramatically. The change in speed and torque is much greater in a series motor than the same change in load imposed on a shunt motor. If the mechanical load is removed entirely, the motor speed increases to dangerous levels, even to the point of destroying the motor by its own centrifugal force. For this reason, series motors are permanently connected to the mechanical load and should not be run with full voltage without a load.

When a mechanical load is added to the series motor, the opposite effect is true. The motor slows with a new, higher torque requirement. As the motor slows, the CEMF in the rotor is reduced and the line current is allowed to increase. This affects both the stator field and the rotor field. With a drop in speed, the torque is increased dramatically, and the motor runs at a lower speed but with much larger torque. This gives the speed torque curve its flattened slope (see Figure 2–2). Note that the series motor has more torque with lower armature current than the other DC motors.

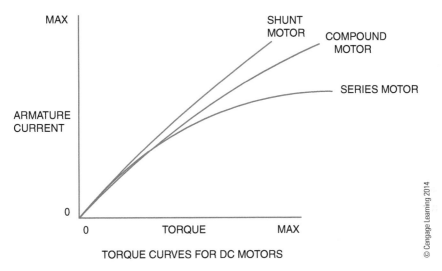

FIGURE 2–2 Comparison of torque versus armature current in DC motors.

If the maximum branch-circuit fuse size (time delay and nontime delay fuses, and inverse time breakers) for any DC motor is limited to 150% of the full-load running current of the motor, the starters used with such motors must limit the starting current to 150% of the full-load current rating. Instantaneous trip breakers allow the rating to be 250% of the rated current, according to *Table 430.52* of the 2011 *National Electrical Code®*. Such starters must be equipped with an automatic, no-load release to prevent the armature from reaching dangerous speeds. The no-load release is set to open the circuit at the armature current corresponding to the maximum speed rating.

The speed of a series motor is controlled by varying the applied voltage. A series motor controller usually is designed to start, stop, reverse, and regulate the speed.

ROTATION

The direction of rotation may be reversed by changing the direction of the current either in the series field or the armature (Figure 2–3).

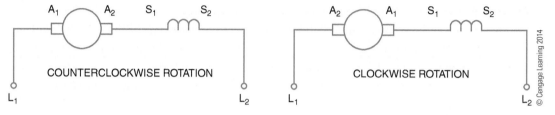

COUNTERCLOCKWISE ROTATION   CLOCKWISE ROTATION

© Cengage Learning 2014

FIGURE 2–3 Standard connections for series motors.

MOTOR RATINGS

Series DC motors are rated for voltage, current, horsepower, and maximum speed.

SUMMARY

The DC series motor has very high starting torque at very low speed (Figure 2–4). This characteristic makes it ideal for traction motors. These motors are used in forklifts and diesel electric locomotives. The relative speed of the motor is controlled by adjusting the applied voltage to both the series field and the armature. The motor can be reversed by changing the direction of current in either the series field or the armature.

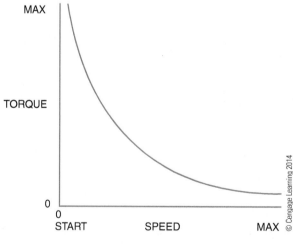

FIGURE 2–4 Starting torque for a DC series motor.

ACHIEVEMENT REVIEW

Select the correct answer for each of the following statements, and place the corresponding letter in the space provided.

1. The torque of a series motor _____
   a. is lower in its starting value than the starting torque for a shunt motor of the same horsepower rating.
   b. depends on the flux of the armature only.
   c. increases directly as the square of the current increases.
   d. increases with a load increase but causes less of a reduction in speed than a shunt motor for the same current increase.

2. For a series motor, _____
   a. the field is operated below saturation.
   b. an increase in both the armature current and the field current occurs because of an increase in load.
   c. the reduction in speed due to an increase in load is greater than in the shunt motor.
   d. all of the above are true.

3. Because a DC series motor has poor speed regulation, _____
   a. a reduction in load causes an increase of current in both the field and armature.
   b. the removal of the mechanical load causes the speed to increase without limit, resulting in the destruction of the armature.
   c. it should not be connected permanently to its load.
   d. it does not require speed control.

4. The speed control for a DC series motor _____
   a. is accomplished using a diverter rheostat across the series field.
   b. has an automatic no-field release feature included on all starters regardless of the limitations on the starting current.
   c. varies with the applied voltage.
   d. is all of the above.

5. A series motor controller usually is designed for _____
   a. cranes.
   b. railway propulsion.
   c. starting heavy loads.
   d. all of the above.

6. Complete the electrical connections for a simple series motor in Figure 2–5.

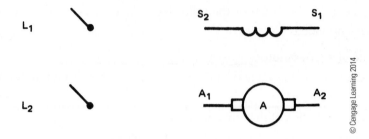

FIGURE 2–5 Series motor connection diagram for question 6.

UNIT

DC COMPOUND MOTORS

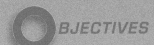BJECTIVES

After studying this unit, the student should be able to

- describe the torque, speed, rotation, speed regulation, and control characteristics of a cumulative compound-wound DC motor.

- perform the preliminary test for the proper installation of a cumulative compound motor.

- connect DC compound motors.

- describe the characteristics of a differential compound-wound DC motor.

- describe the characteristics of a cumulative compound-wound DC motor.

Compound-wound motors are used whenever it is necessary to obtain speed regulation and torque characteristics not obtainable with either a shunt or a series motor. Because many drives need a fairly high starting torque and a constant speed under load, the compound wound motor is suitable for these applications. Some of the industrial applications include drives for passenger and freight elevators, stamping presses, rolling mills, and metal shears.

The compound motor has a normal shunt winding and a series winding on each field pole. As in the compound-wound DC generator, the series and shunt windings may be connected in long shunt, as shown in Figure 3–1(A), or short shunt, shown in Figure 3–1(B).

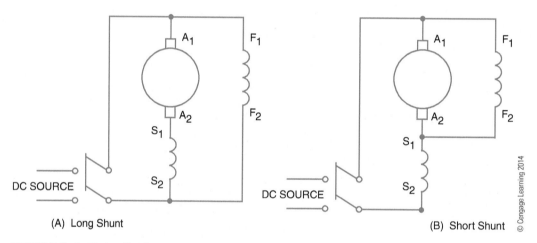

(A) Long Shunt   (B) Short Shunt   © Cengage Learning 2014

FIGURE 3–1 Motor field connections.

When the series winding field is connected to aid the shunt winding field, the machine is a *cumulative compound motor*. When the series field opposes the shunt field, the machine is a *differential compound motor*. Using Fleming's left-hand rule for electromagnets, it can be seen that the two windings will either reinforce each other or try to cancel each other (Figure 3–2).

TORQUE

The operating characteristics of a cumulative compound-wound motor are a combination of those of the series motor and the shunt motor. When a load is applied, the increasing current through the series winding increases the field flux. As a result, the torque for a given current is greater than it would be for a shunt motor. However, this flux increase causes the speed to decrease to a lower value than in a shunt motor. A cumulative compound-wound motor develops a high torque with any sudden increase of load. It is best suited for operating varying load machines such as punch presses.

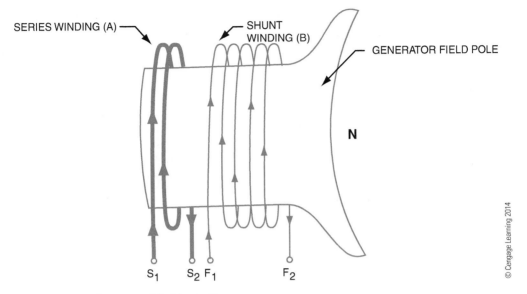

SERIES WINDING (A)

SHUNT
WINDING (B)

GENERATOR FIELD POLE

N

S_1 S_2 F_1 F_2

© Cengage Learning 2014

FIGURE 3-2 Compound field windings.

SPEED

Unlike a series motor, the cumulative compound motor has a definite no-load speed and will not build up to destructive speeds if the load is removed.

Speed Control

The speed of a cumulative compound motor can be controlled by using resistors in the armature circuit to reduce the applied voltage. When the motor will be used for installations where the rotation must be frequently reversed, such as in elevators, hoists, and railways, the controller should have voltage dropping resistors and switching arrangements to accomplish reversal. This was the general mode of control for older installations of DC motors. Now, most of the controllers use electronics to accomplish the same effect.

Electronic Speed Control

A block diagram approach to the electronic speed control of a DC motor is presented in Figure 3–3. The AC line is rectified by a full-wave bridge to supply pulsating DC to the shunt field at a steady value. The input is shown as a single-phase AC source, but the input may be a three-phase AC source, especially for larger motors. The DC supplied to the armature and the series field is controlled by a silicon-controlled rectifier (SCR) or other solid-state electronics such as a MOSFET (metal-oxide semiconductor field-effect transistor). By adjusting the firing time of the semiconductor, either more or less of the DC voltage available can be applied to the

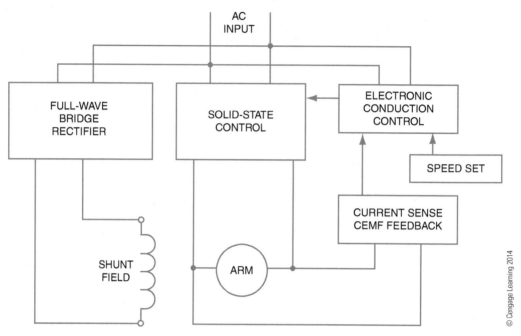

FIGURE 3–3 Block diagram of electronic speed control for a DC motor.

armature. If a low amount of DC voltage is applied, the torque is low and the resultant speed is low. If the solid-state device is fired early in the waveform and allowed to conduct for most of the cycle, then a larger amount of voltage is applied to the armature, more current results, and torque increases to spin the armature at a higher speed. Speeds below rated motor speed are produced by lowering the amount of voltage applied to the armature while keeping the shunt field steady; if higher than normal speed is needed, the shunt field can be weakened. Most speed controls have some type of feedback to sense the armature current, compare it to the set speed, and adjust the electronic conduction control to compensate for varying mechanical loads to keep the motor speed regulated. (For more details, see Unit 6.)

Speed Regulation

The speed regulation of a cumulative compound-wound motor is inferior to that of a shunt motor and superior to that of a series motor. It is a compromise between a series motor and a shunt motor, as seen in Figure 3–4.

The graph in Figure 3–4 shows that the percent of speed regulation of a compound-wound DC motor is lower than that of a shunt motor but higher than that of a series motor.

The speed versus torque curves presented in Figure 2–2 show the relationship between the shunt, series, and compound motor. The series has the highest torque at the lowest speed, but

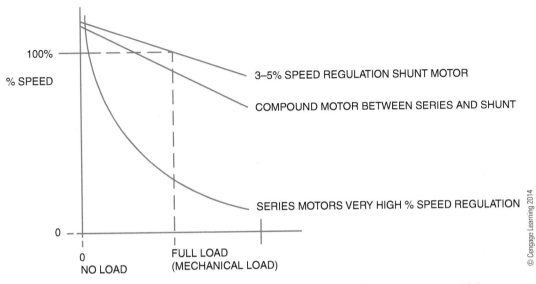

FIGURE 3-4 Graph of speed regulation of shunt, compound, and series-connected DC motors.

the speed regulation is very poor. The shunt motor has less torque at higher loads, but the speed regulation is very good. The compound motor has characteristics of both, in that the torque and speed regulation characteristics are in between the shunt and the series motor.

ROTATION

The rotation of a compound-wound motor can be reversed by changing the direction of the current in the field circuits or the armature circuit (Figure 3–5). Because the series field coils must also be reversed if the shunt field is reversed, it is conventional to reverse the current in the armature only.

PRELIMINARY TEST FOR CUMULATIVE COMPOUNDING

When a motor is first connected, it is important to determine the continuity of the shunt field circuit. In addition, for a

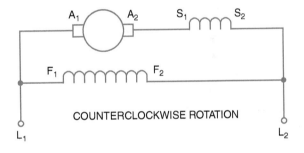

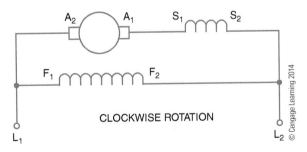

FIGURE 3–5 Standard connections for compound motors.

compound-wound motor, the proper magnetic polarity of the shunt and series field must be determined. Standardized tests determine these conditions. For example, when a compound motor is ready to be placed into operation, a double check for proper compounding is necessary. It must be verified that the series field is aiding the shunt field flux. Connecting the compound motor is referred to as *cumulative compounding*. If this is not done, the motor may start at light load under the influence of the shunt field. As the load increases, the series field becomes more powerful. If the series field is an opposite polarity from the shunt field, the stator field becomes weaker. Eventually the series field becomes more dominant, and it may suddenly reverse the direction of rotation of the motor.

The following test verifies field polarity. With no load on the motor, connect the motor as a shunt motor only, without placing the series field in the circuit. Momentarily start the motor and note the direction of rotation. Keep track of which motor lead markings are connected to which line leads. Now connect the motor as a series motor, without the shunt field. Momentarily start the motor. *Do not* let it run without a mechanical load. Again, be sure to note which series field leads are connected to which line leads. If the motor turns the same direction each time, reconnect the shunt field to the original line leads. If not, reverse the shunt field leads so that both series and shunt motor connections cause the same direction of rotation. This establishes cumulative compounding.

Differential Compounding

Almost 0% speed regulation can be obtained with a differentially compounded motor. When a motor is connected as a differential compound machine, the series field opposes the shunt field so that the field flux is decreased as a load is applied (Figure 3–6). As a result, the speed remains

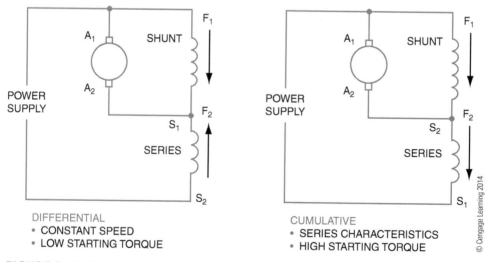

FIGURE 3–6 Magnetic polarities of compound motors.

substantially constant with an increase in load. With overcompounding, a slight increase in speed is possible with an increase in load. This speed characteristic is achieved only with a loss in the rate at which torque increases with load.

Because the field decreases with a load increase, a differential compound motor has a tendency for load instability. When starting a differential motor, it is recommended that the series field be shorted because the great starting current in this field may overbalance the shunt field and cause the motor to start in the opposite direction.

A differential machine is connected and tested on installation using the same procedure outlined for a cumulative compound motor. For the differential motor, however, the series windings should be connected in the opposite direction from that of the shunt winding. Do not exceed the load on the nameplate, or reversal of direction may occur as noted in the description of cumulative compounding.

SUMMARY

The DC compound motor is used where a compromise is needed between the series and the shunt motor. The compound motor has better speed characteristics than the series motor and better torque characteristics than the shunt motor. The motor can be connected so that the shunt field and the series fields are in the same direction. This is the *cumulative connection*. You should perform a test to be sure the motor is connected as intended. The cumulative connection reacts differently than the differential connection. Many DC motors are now driven by electronic drives. The same basic concepts are used to control speed and direction.

ACHIEVEMENT REVIEW

1. Circle the letter for each of the following statements that applies to a cumulative compound-wound DC motor.
   a. The speed regulation of a cumulative compound-wound DC motor is better than that of a shunt motor.
   b. The speed of a cumulative motor has a no-load limit.
   c. The speed of the motor decreases more for a given increase in load than does a differential motor.
   d. A cumulative motor has less torque than a shunt motor of the same hp rating for a given increase in armature current.
   e. The speed regulation of a cumulative motor is better than that of a series motor.
   f. A cumulative motor develops a high torque with a sudden increase of load.
   g. To reverse the direction of rotation, the current in either the armature or the shunt field must be reversed.

h. A cumulative motor is connected so that the series flux aids the shunt winding flux.

i. When installing a cumulative compound-wound motor, the direction of rotation should be the same when testing the motor for operation either as a series motor or a shunt motor.

2. Circle the letter for each of the following statements that applies to a differential compound-wound DC motor.

a. A differential motor is used in applications where an essentially constant speed at various loads is required.

b. The starting torque for a differential motor is higher than that of a cumulative motor.

c. The motor may reverse its direction of rotation if started under a heavy load.

d. This motor develops a speed instability because the flux field decreases with a load increase.

e. When starting a differential motor, the shunt field should be shorted because of the great starting current.

3. Arrange the following steps numerically in the correct sequence to test for the proper connections to operate a cumulative compound-wound motor. Place the step number in the space provided, starting with number 1.

_____ a. If rotation is in direction opposite to that desired, reverse the series field leads.

_____ b. Start motor as a shunt motor and observe rotation.

_____ c. Connect only the series field, start motor, and immediately shut it down while noting direction of rotation.

_____ d. Connect the shunt field circuit; the motor is now ready for operation.

UNIT

SUMMARY REVIEW
OF UNITS 1–3

4

OBJECTIVES

- To provide the student with an opportunity to evaluate the knowledge and understanding acquired in the study of the previous three units.

Select the correct answer for each of the following statements, and place the corresponding letter in the space provided.

1. The twisting effect of a motor shaft is called its _____
   a. turning power.
   b. horsepower.
   c. RPM.
   d. torque.

2. The twisting effect of a DC motor is produced primarily by _____
   a. the armature.
   b. the rotor.
   c. a current-carrying conductor in a magnetic field.
   d. torque in the field coils.

3. A DC motor is required to maintain the same speed at no load and full load. This type of operation can only be obtained by using a _____
   a. series motor.
   b. shunt motor.
   c. differential compound-wound motor.
   d. cumulative compound-wound motor.

4. As a load is applied to a DC shunt motor the _____
   a. field current increases.
   b. CEMF increases.
   c. armature current increases.
   d. torque developed decreases.

5. The speed of a DC shunt motor _____
   a. increases with an increase in load.
   b. decreases with an increase in applied voltage.
   c. decreases if the field strength is increased.
   d. decreases less than a series motor of the same hp for the same increase in load.

6. As load is applied to a DC series motor the _____
   a. field current decreases.
   b. field voltage increases.
   c. armature current decreases.
   d. armature voltage increases.

7. The load requirements of a particular DC motor installation necessitate extremely high starting torque. If speed regulation is not important, use a _____
   a. series motor.
   b. shunt motor.
   c. differential compound-wound motor.
   d. cumulative compound-wound motor.

8. As a load is applied to a cumulative compound-wound DC motor its _____
   a. speed decreases.
   b. CEMF decreases.
   c. torque decreases.
   d. series field current decreases.

9. In a cumulative compound-wound DC motor, the _____
   a. series winding develops the major part of the total flux.
   b. series and shunt windings develop field flux in the same direction.
   c. shunt winding must be connected across the brushes.
   d. series windings do not pass the shunt field current.

10. Circle the letter of the statement that is not true for a DC shunt motor.
    a. Torque is proportional to the field current.
    b. The same voltage is applied to armature and field circuits.
    c. The no-load speed is controlled by the impressed voltage.
    d. The motor is suitable for installations requiring substantially constant speed with variable loading.

11. In a differential compound-wound DC motor _____
    a. the series and shunt fields establish flux in the same direction.
    b. the series winding acts to reduce speed as load is applied.
    c. an increase in total current input as the result of loading increases the shunt field current.
    d. changes in torque result in change of current in the series field windings.

12. If the direction of field flux and the direction of armature current are changed, the torque developed by the motor is _____
    a. stronger.
    b. less.
    c. the same.
    d. reversed.

13. A generator shunt field winding is _____
   a. high resistance.
   b. low resistance.
   c. noninductively wound.
   d. embedded in the armature.

14. For proper operation in a four-lead DC motor, leads S_1 and S_2 should be connected to A_1 and A_2 in _____
   a. parallel.
   b. series.
   c. shunt.
   d. series parallel.

15. Decreasing the resistance of a generator field rheostat _____
   a. decreases the flux.
   b. decreases the voltage.
   c. increases the voltage.
   d. decreases the speed.

16. A series field, if connected across a motor armature and energized, _____
   a. makes the motor race dangerously.
   b. causes a short circuit.
   c. creates excessive flux.
   d. acts as a shunt field.

UNIT 5

STARTING AND OPERATING DC MOTORS

OBJECTIVES

After studying this unit, the student should be able to

- describe the need for starting controls for DC motors.
- define manual starters used with DC motors.
- describe the two basic types of manual DC motor starters.
- identify the drum controller section and read connection diagrams.
- describe the connections that occur at each first position in forward and reverse of a drum controller.

Manual starters for DC motors are rare because electronic controls are much more efficient and practical. A short review of the manual methods called three-point and four-point starters may be useful, as some electricians may still need to maintain these starting methods.

Two factors limit the amount of current used by a motor that is supplied from a DC source: (1) the counter-electromotor force (CEMF) produced in the rotor by the spinning of the conductors through a magnetic field and (2) the actual resistance of the motor windings. The wire resistance is typically very low, less than 10 ohms, and the rotor has no CEMF when the rotor is at a standstill or just starting. Therefore, the current is very high at the starting of the DC motor. The same effect is also seen in AC motors, but the introduction of impedance of the motor coils due to AC helps limit the starting current. In DC motors, the need for a starting controller with an in-line resistor or variable rheostat for limiting the starting current was devised. This starting rheostat, or DC motor starter, is described by NEMA (National Electrical Manufacturers Association) as a device used to accelerate the motor to its rated speed in one direction and limit the current to the armature circuit to a safe value during starting and accelerating. Two styles of manual starters were used.

THREE-TERMINAL STARTER FOR DC MOTORS

A three-terminal starter is shown in Figure 5–1, connected to a shunt-wound motor. Note that the tapped resistance is connected in series with the armature, yet the high resistance field receives full voltage immediately upon starting. The idea is to get the field flux at full value

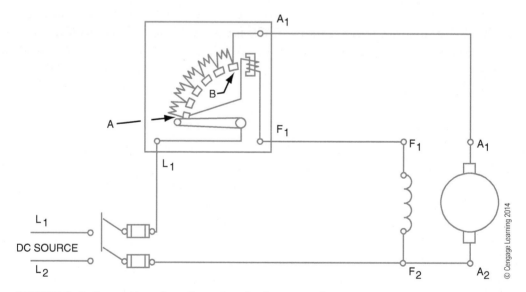

FIGURE 5–1 Connections for a three-terminal starting rheostat.

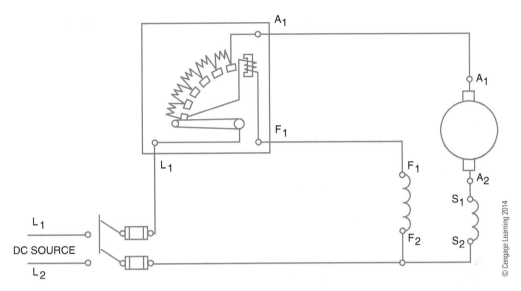

FIGURE 5–2 Three-terminal starting rheostat connected to a cumulative compound-wound motor.

immediately so the armature immediately begins producing CEMF as it begins spinning. As the starter is moved from left to right, portions of the armature series resistance are shorted out, and eventually there is no more resistance in series with the armature. The electromagnet on the far right actually holds the arm in the "run" position. If the shunt field has an open circuit at any time, the arm releases the controller arm, causing the motor to stop as a precaution. This "no field" circuit is necessary to prevent a high-speed runaway of the DC shunt motor at no load. In Figure 5–2, a compound motor is connected to a three-terminal starter. The only change is the addition of the series field to the diagram.

FOUR-TERMINAL STARTERS FOR DC MOTORS

This slight variation on the three-terminal starter is used if there is a need for variable speed control using a field rheostat (see Figure 5–3). As the arm is raised to the first position, the shunt field receives full voltage, yet there is still resistance in the armature circuit. As the arm is moved to the right, resistance is added to the shunt field circuit but removed from the armature circuit. Because adding resistance to the shunt field circuit reduces the current to the field, the holding electromagnet is not in series with the shunt field but is connected across the line, hence the fourth terminal. This process makes this starter more like an AC starter with "no-voltage" release. If line voltage drops too low, the electromagnet releases the arm, which springs back to stop the motor.

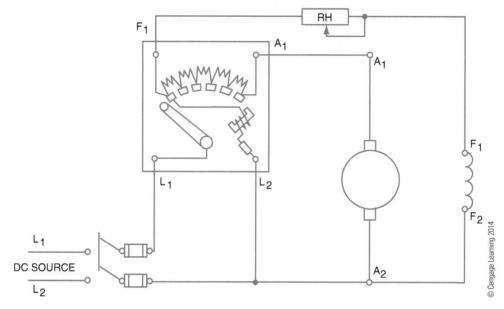

FIGURE 5-3 Connections for a four-terminal starting rheostat.

NATIONAL ELECTRICAL CODE® FOR MOTOR STARTERS

Part VII in *Article 430.82(C)* refers to the controller design and the need for starting rheostat controls. Note that *430.89* requires speed limitations on separately excited DC motors and on series motors as well as motor generators and converters that can be driven by DC at excessive speed. Remember that DC series motors that can run at no load may generate runaway speeds and literally throw themselves apart.

DRUM CONTROLLERS

Drum controllers are used when an operator is controlling the motor directly. The drum controller is used to start, stop, reverse, and vary the speed of a motor. This type of controller was used on crane motors, elevators, machine tools, and other applications in heavy industry. As a result, the drum controller had to be more rugged than the starting rheostat.

A drum controller with its cover removed is shown in Figure 5–4. The switch consists of a series of contacts mounted on a movable cylinder. The contacts, which are insulated from the cylinder and from one another, are called *movable contacts*. Another set of contacts, called *stationary contacts,* is located inside the controller. These stationary contacts are arranged to touch the movable contacts as the cylinder is rotated. A handle, keyed to the shaft for the movable cylinder and contacts, is located on top of the drum controller. This handle can be moved either clockwise or counterclockwise to provide a range of speed control in either direction of rotation. The handle can remain stationary in either the forward or reverse direction due to a roller and a notched wheel. A spring forces the roller into one of the notches at each successive position of the controller handle to keep the cylinder and movable contacts stationary until the handle is moved by the operator.

A drum controller with two steps of resistance is shown in Figure 5–5. The contacts are represented in a flat position in this schematic diagram to make it easier to trace the circuit connections. To operate the motor in the forward direction, the set of contacts on the right must make contact with the center stationary contacts. Operation in the reverse direction requires that the set of movable contacts on the left make contact with the center stationary contacts.

Note in Figure 5–5 that the controller handle can be set to three forward positions and three reverse positions. In the first forward position, all the resistance is in series with the armature. The circuit path for the first forward position is as follows:

FIGURE 5–4 Drum type of controller showing contact fingers.

1. Movable fingers a, b, c, and d contact the stationary contacts 7, 5, 4, and 3.

2. The current path is from the negative side of the line to contact 7, from 7 to a, from a to b, from b to 5, and then to armature terminal A_1.

3. After passing through the armature winding to terminal A_2, the current path is to stationary contact 6, and then to stationary contact 4.

4. From contact 4, the current path is to contact c, to d, and then to contact 3.

5. The current path then goes through the armature resistor, to the series field, and then back to the positive side of the line.

The shunt field of the compound motor is connected across the source voltage. On the second forward position of the controller handle, part of the resistance is cut out. The third forward position cuts out all the resistance and puts the armature circuit directly across the source voltage.

In the first reverse position, all the resistance is inserted in series with the armature.

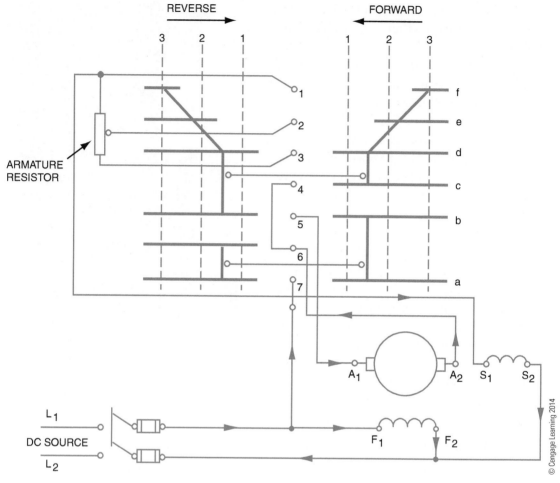

FIGURE 5–5 Schematic diagram of a drum controller connected to a compound-wound motor.

Figure 5–6 shows the first position of the controller in the reverse direction. The current in the armature circuit is reversed. However, the current direction in the shunt and series fields is the same as the direction for the forward positions. Remember that an earlier unit showed that a change in current direction in the armature *only* resulted in a change in the direction of rotation.

The second reverse position cuts out part of the resistance circuit. The third reverse position cuts out all the resistance and puts the armature circuit directly across the source. Drum controllers with more positions for a greater control of speed can be obtained. However, these controllers all use the same type of circuit arrangement shown in this unit.

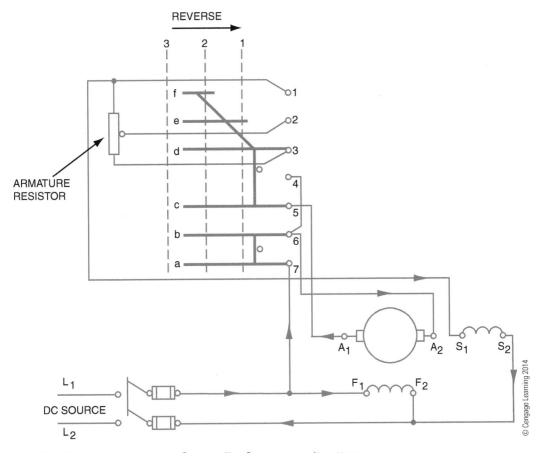

FIGURE 5-6 First positions of controller for reverse directions.

SUMMARY

Three-terminal and four-terminal starting rheostats are not used much any more. The advent of electronic starters has all but replaced the mechanical starters. The concept of the starter is still used, and the safeguards for motor operation are still important.

DC series motors require a different starting controller than a shunt or compound motor. The holding circuit for the controller is in series with the starting resistance. If there is a low-voltage or no-voltage condition, the starter is returned to the off position. Drum controllers are still frequently used with AC as well as DC motors. It is important to be able to read the connection diagrams and the sequence diagrams on drum-type controllers.

ACHIEVEMENT REVIEW

1. What are the two functions of a motor starter?

    a. _____

    b. _____

2. Show the connections of a three-terminal starting rheostat to a shunt motor.

3. Show the connections of a four-terminal starting rheostat to a shunt motor.

4. Complete the connections in Figure 5–7 to show that the cumulative compound-wound motor can be started from the four-terminal starting rheostat. Also connect the field rheostat in the circuit for above-normal speed control.

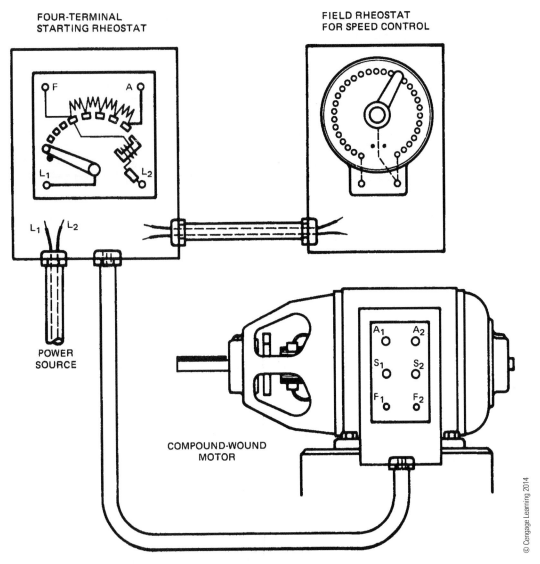

FOUR-TERMINAL
STARTING RHEOSTAT

FIELD RHEOSTAT
FOR SPEED CONTROL

POWER
SOURCE

COMPOUND-WOUND
MOTOR

© Cengage Learning 2014

FIGURE 5–7 Connection diagram for question 4.

5. Why is a drum controller used in many industrial applications? _____

_____

_____

6. A series starter with no-load protection is used to prevent the series motor from reaching _____ at low loads.

7. A drum controller provides the following types of control for a DC motor:

_____

_____

_____

UNIT 6

ELECTRONIC DC MOTOR CONTROL

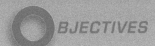 BJECTIVES

After studying this unit, the student should be able to

- identify the major sections of an electronic DC motor drive.

- determine the operating features available to the end user.

- describe acceleration and deceleration and braking techniques.

- explain the operating concepts of DC drives for different motors.

DC motors are used in a variety of applications for exact speed control and for high torque requirements that have not readily been available with AC motors. The DC motor and drive system has been available for some time, and many industrial installations are dependent on the operating characteristics of the motor and associated drive system. Until the more recent introduction of the variable frequency drive (VFD), and especially the vector drive, precise speed control and acceleration, deceleration, and electrical braking were not available with AC motors.

DC DRIVE BASICS

The DC controller is based on the same principles as the older electromechanical systems for controlling DC motors. In other words, to control the speed, the current is reduced to the armature of the DC motor to reduce the speed on the motor below normal full operating design speed, and the main field is weakened for speeds above normal speed. The difference between the older starters and the electronic one is efficiency and much tighter control. The old starters inserted resistance into the armature circuit to reduce the voltage and current to the armature and therefore dissipate the energy in the resistors and waste the power, making it inefficient. There was little feedback to the controller, so there was no adjustable process to control the exact speed of the motor. With electronics, there is little wasted energy and much more precise control of actual motor operations.

DC DRIVE BLOCK CIRCUITS

The DC motor requires the application of direct current to the actual motor windings. The DC drive controller will change, or rectify, the available AC to DC and adjust the DC to control the motor operations. Two general schemes are used to deliver the DC to the motor. Each of the methods provides the same basic functions. See Figure 6–1 for the block diagram of a controller. One block provides DC to the field, and one block supplies controlled DC to the armature. A set point control adjusts the firing of output silicon- controlled rectifiers (SCRs or triodes for alternating current [TRIACs]). Feedback circuits tell the controller what the motor is doing.

SIMPLE CONTROL

One method used for small motors with only basic speed and torque requirements is to use only half of the incoming AC, rectify it with a half-wave rectifier system, and deliver half-wave DC to the motor armature (see Figure 6–2). In most cases, the shunt field is supplied by a full-wave bridge circuit that supplies rectified DC to the motor field. This can be an adjustable voltage also if the motor has to operate above rated speed. The armature power is typically controlled by an SCR that is turned on at different points in the waveform to adjust the voltage to the motor armature to control the speed. In the case of a motor where exact speed control and high torque are not critical, the half-wave controller may be sufficient and most

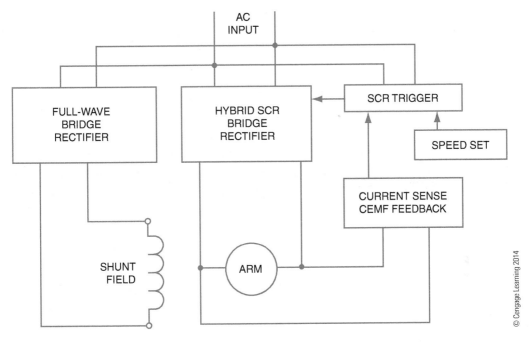

FIGURE 6-1 Block diagram of electronic controller used for a shunt motor.

economical. The firing of the SCR or the gate control may be of the *amplitude control* or the *phase control* type. The amplitude control only has control over the first half of the sinewave, as seen in Figure 6–3. This is the least costly control scheme and provides the least amount of control. The more common type is the phase control method that allows the half-wave almost 180 electrical degrees of control (see Figure 6–4).

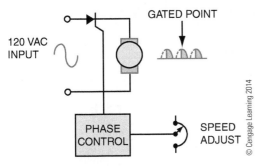

FIGURE 6-2 Phase control controls output for nearly 180°.

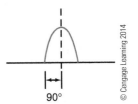

FIGURE 6-3 Amplitude control has a control range for the first 90° of half-sinewave (0 to 90° control).

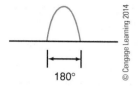

FIGURE 6–4 Phase control has control for nearly the full 180° of the half-sinewave (0 to 180° control).

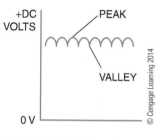

FIGURE 6–5 DC voltage with ripple content.

DC POWER SUPPLIES

In almost all situations currently encountered, the DC voltage sources needed to supply motors and other electrical apparatus are produced by solid-state rectifiers that convert alternating current (AC) to direct current (DC). There are several different methods used to rectify the AC and create various levels of DC output from the rectifier circuits at different degrees of smoothed (low-ripple) DC. As in DC generators where the DC is produced through commutation of a generated AC wave, the commutated DC is not always a smooth DC voltage as you would see from a battery output. Electronic rectifier circuits also produce a DC with different percentages of ripple to the DC voltage. The small-ripple value of the DC voltage can be seen in Figure 6–5. As you can see, the voltage output is DC, meaning that the current always flows in one specific direction, but it flows at unequal amounts. There are, in fact, a peak and a valley to the DC voltage. This fluctuating DC is sometimes referred to as the AC component (fluctuating component) of the DC wave.

The rectifier circuit can be as simple as a single diode used with a single-sinewave AC input, as in Figure 6–6. The output waveform is DC half-wave rectification. This means that only half of the AC voltage is used and the other half of the AC input voltage is blocked, or not used. The DC value of the output voltage is approximately 0.314 (or π) times the peak AC value of the input. If the peak of the AC input is 100 volts, a meter set to DC on the output would measure approximately 31.4 VDC. If you look at the DC with an oscilloscope, it appears very bumpy; in other words, it has a lot of ripple. When you measure the AC (fluctuating component) of the waveform using a meter that blocks the DC level, you measure near the value calculated by AC peak × 0.385, or about 38.5 volts in this example. The percent ripple is calculated by the formula

% ripple = AC ripple component/DC value

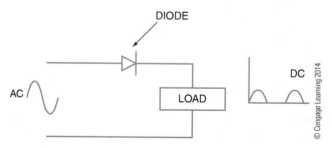

FIGURE 6–6 One-half wave rectifier uses a single diode.

Compare the DC value to the AC value for a half-wave rectifier and you get approximately 121% ripple. We discuss how to filter the DC to reduce the ripple at a later time.

Another rectifier circuit that provides more DC for a specific AC input is known as the full-wave rectifier circuit. This circuit requires two diodes and the use of a center-tapped secondary AC transformer. By viewing Figure 6–7, known as a center-tapped, full-wave rectifier circuit, you see that the full input voltage is used in the DC output. In this circuit both halves of the AC input waveform are used to supply the DC load. For a half cycle of input voltage, the top of the circuit is the most positive part of the circuit, and current flows from the center tap of the transformer, through the load, through diode 1, back to the source. In the other half cycle, the current again flows from the center tap through the load, through diode 2, back to the bottom side of the source transformer, which is now the most positive point of the circuit. In this circuit the DC value of the output voltage is twice the value of the half-wave rectifier. The output is now approximately 0.637 (2π) times the peak of the AC input voltage. The AC ripple content has decreased to AC peak × 0.305. There is less nonconducting time for the DC voltage, and the ripple content has decreased. In fact, with a higher DC content and a lower fluctuation, or AC component, the ripple content has decreased to approximately 48%. With less ripple, less filtering is needed to create smooth DC voltage output.

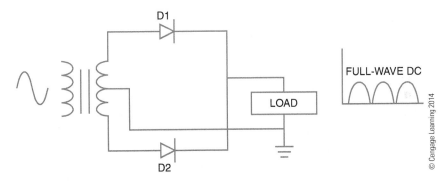

FIGURE 6–7 Center-tapped, full-wave rectifier circuit.

Yet another type of rectifier circuit is the full-wave bridge circuit, as in Figure 6–8. As the name implies, the DC uses the full wave of the AC input. The bridge refers to the style of circuit converting AC to DC. With the use of four relatively inexpensive diodes, the expensive center-tapped transformer can be eliminated. See Figure 6–9 for each half-cycle current path through the bridge of diodes. The output of the bridge has the same values of DC as the center-tapped, full-wave rectifier if you ignore the small diode voltage drops of the rectifiers. The waveform of the DC output has a frequency as well as a ripple content. The ripple frequency is determined by the number of positive peaks that occur in 1 second. In both full-wave rectifier circuits, the number of positive peaks that occurs is twice that of the AC input-positive peaks, as verified in

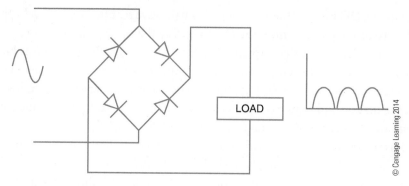

FIGURE 6-8 Full-wave bridge rectifier circuit.

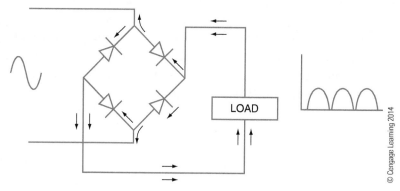

FIGURE 6-9 Full-wave bridge with current directions for each half cycle.

Figure 6–12. If the input AC waveform is at a standard 60 Hz, or 60 positive peaks per second, the full-wave DC has 120 positive peaks for a ripple frequency of 120 Hz. By comparison, in the half-wave rectifier the input frequency is equal to the output DC frequency. Remember the AC component does not reverse the direction of current flow but only changes the value of the DC. The frequency relates to the AC component of the DC output.

In DC motor control, the fields of the DC motors are often supplied by a steady DC value provided by diode rectifier circuits. To smooth out the DC level and make it more consistent, the ripple effect of the DC has to be filtered or smoothed. With a higher ripple frequency and smaller ripple content, as in full-wave rectification, the filtering can be done effectively and efficiently with capacitor filters.

Capacitive filtering uses a capacitor or set of capacitors in parallel with the load. The circuit in Figure 6–10 shows that the capacitor charges up to the peak of the waveform as current flows through the load and also charges the capacitor. In the half cycle, when the supply current path is blocked, the capacitor begins to discharge through the load in the direction

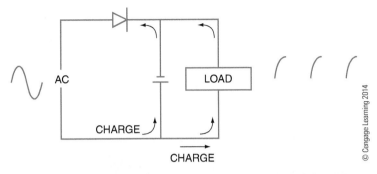

FIGURE 6-10 Capacitive filter with charge current directions.

shown in Figure 6–11. The effect is to fill in the gap of nonconducting diode time and reduce the ripple and increase the DC value of voltage. By using the same concept on the full-wave bridge rectifier circuit of Figure 6–12, the DC value becomes smoother and the ripple becomes less objectionable. With the load consisting of the motor coils producing the magnetic field, the steady DC makes the motor performance more reliable and predictable.

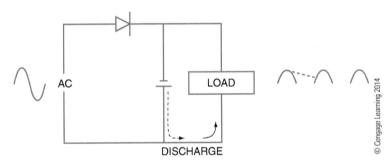

FIGURE 6-11 Capacitive filter with discharge current directions.

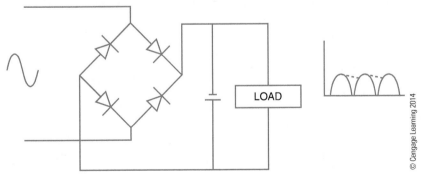

FIGURE 6-12 Full-wave bridge needs less filtering than a half-wave rectifier.

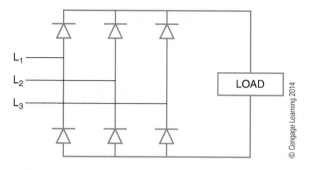

FIGURE 6-13 Three-phase rectifier circuit.

Most DC electronic speed controls use a separate source to provide steady DC to the fields. These supplies are made of full-wave bridges or more typically use a three-phase rectifier system, as in Figure 6–13. In this system, all three phases are rectified, and the DC level remains high and the ripple percentage is very low. This means the DC supply needs very little filtering.

The DC supplies that are used to provide the DC for the armature are constructed a little differently. The reason for the difference is the method used to control the speed of a DC motor. For speeds below full speed, the method used reduces the voltage applied to the armature while leaving the field at full strength. Therefore, it is necessary to create a variable voltage DC supply to control the armature power. For most DC controllers, the primary power tends to be three-phase AC. The schematic diagram in Figure 6–14 is used with the addition of a control network that allows the thyristors, such as TRIACs, to be turned on at various times. Essentially, if the TRIACs are gated at various angles, the output DC can be controlled rather precisely. If the TRIACs are turned on at 0°, the output voltage is very high positive, based on the formula

$$V_{DC} = 1.35 \times AC_{RMS} \times \text{cosine of the firing angle}$$

With 460 VAC applied and a 0° firing angle, the DC is approximately 620 volts. At a 90° firing angle, the DC voltage is zero because there is just as much positive as negative waveform, and at a 180° firing angle the DC is all negative at −621 volts. Angles between 0° and 90° yield DC between 621 and 0 VDC. This process allows us to control the armature voltage that adjusts the speed of the DC motor.

As seen in Figure 6–14, there is a large diode across the armature connection called a *flywheel diode, freewheeling diode,* or *kickback diode.* The diode is connected in the control so that

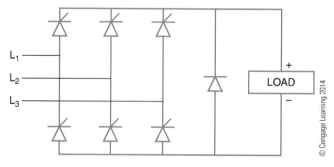

FIGURE 6-14 Thyristors used to control the DC voltage output from a three-phase source.

it is reversed biased, blocking current flow, while the normal voltage is applied to the armature. When the supply DC is turned off, the magnetic field in the armature collapses and creates a large inductive kick voltage, opposite to the applied voltage. This large inductive kick voltage could be enough to break down the motor windings or the motor controller; therefore, the fly-wheel diode is forward biased by the inductive kick, and the kick is reduced by allowing current to flow back through the armature windings, which slows the magnetic field collapse.

SAFETY CIRCUITS

In most controllers, a safety circuit monitors the DC field. If this stator field is lost, the motor may accelerate to dangerous speeds, especially with no load. If the field is lost on compound motors, the motor becomes a series motor and operates as a series motor would. The power to the armature must be disconnected. The current to the field can be monitored by placing a low-ohm resistor in series with the field. Current flowing to the field passes through the resistor and creates a voltage drop. The voltage drop becomes part of the control circuit for the armature. If there is no field sensor voltage, the SCR firing circuit for the armature shuts down. Another method is to use a DC current flow sensor called a *Hall effect sensor,* which is essentially a current transformer specifically designed to sense DC current flow. Again, this sensor provides information to the armature control circuit.

Another type of safety is the current control to the armature. Because there is only DC resistance to current flow in the windings when the motor is not spinning, the initial current flow to a stopped or stalled motor may be very large. To avoid damaging the motor or the electronic controller, current sensors can be used to adjust the firing angle of the SCRs to control armature current. One method of sensing the current is the same as the field sensor. By placing a low-ohm resistor in series with the armature current, the voltage dropped across the resistor can be used to operate a control point in the firing circuit. If the current is too high, the voltage drop across the resistor is high, and the signal is sent to the controller to reduce the armature current. Another method is to measure the AC current going into the output SCR output controller. By using AC current transformers, the input current is monitored, which is indicative of the output DC current. When the output is too high, the input monitors send a signal to the controller to reduce the "on" time of the SCRs (see Figure 6–15).

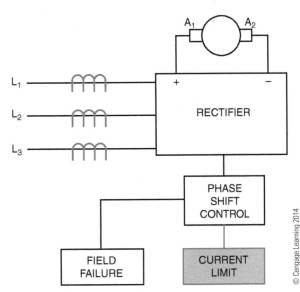

FIGURE 6–15 Current flow to armature is limited.

© Cengage Learning 2014

SPEED CONTROL AND REGULATION

As mentioned earlier, the speed of the DC motor is easily controlled by adjusting the amount of time the thyristors are allowed to conduct by controlling the "on" time during each cycle. If the controller is operating in an *open loop* mode, there is no feedback from the motor to the controller. This means the thyristors will fire as determined by the input control signal or the speed-setting potentiometer. If the motor is to maintain a set speed under varying loads, there must be feedback from the motor to create a *closed loop* system. Often times the actual speed of the motor is monitored by a tachometer attached to the motor. The tachometer output signal is fed back to the firing control to adjust the firing of the thyristor to compensate for load-induced speed changes. Another method to monitor the exact speed of the motor is to use an armature-voltage feedback circuit. This circuit basically measures the exact voltage across the armature. If the motor speed slows down, less CEMF is produced before any change occurs in the control. This reduced CEMF causes the voltage across the armature to be reduced. The reduction in the armature voltage is fed back to the controller, causing the thyristors to be turned on earlier in the cycle and more current to be sent to the armature.

Internal resistance (IR) compensation is another system designed to improve the speed regulation of a DC motor. The IR of the motor is relatively low. The IR compensation monitors the actual current to the armature by using a small series resistor (see Figure 6–16). If the motor speed slows down, less CEMF is produced by the motor. If the firing does not change, more current is allowed to flow to the motor. This IR sensor detects the increased current flow and sends a signal to the thyristor firing circuit to increase the "on" time, thereby increasing motor speed back to the set point, which increases the CEMF back to its original set value. IR compensation is typically one of the parameters set on the controller.

Many controllers have minimum and maximum speed settings. These settings are used to provide for safe operating limits for the motor. The minimum speed is the lowest possible speed,

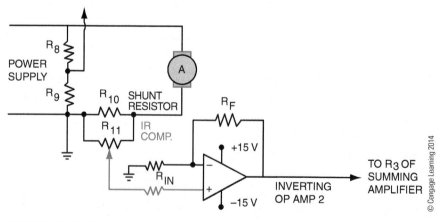

FIGURE 6–16 IR compensation, speed-regulation circuit.

even though the speed control potentiometer is set to zero or the minimum setting. This may be used when a process should not be allowed to completely stop even if the speed is reduced as far as possible. Likewise, the upper end of the speed control is often limited with a maximum speed adjustment. This control sets the maximum allowable speed of the motor even if the controller is capable of higher speeds with the speed-control potentiometer. In addition to the minimum and maximum speed settings, parameters control the acceleration and deceleration of the motors.

Acceleration times can be controlled by setting the ramp-up time from minimum speed to set speed. This feature prevents excessive current draw from the line and reduces electrical stress on the motor and mechanical stress on the driven equipment. Too sudden of an increase can cause the shaft to shear off if the mechanical load is heavy. A safety feature called an *electronic shear pin* is designed to monitor this type of situation. If the current increase to the motor is too swift, it might be an indication of a jammed motor, and the sudden increase in twisting effort might cause the shaft to shear. The electronic shear pin reacts to the amount—and rate—of current increase to the motor. A smooth acceleration is controlled by gradually increasing the "on" time of the SCRs.

Deceleration is the gradual stopping of the motor. This time can be set to bring the motor from set speed to minimum speed. By slowly decreasing the "on" time of the SCRs, the motor slows gradually, thus reducing the mechanical stress on equipment if the motor were to come to an abrupt stop.

Above-normal (or base) speed can also be obtained within the mechanical limits of the motor by adjusting the shunt field of the motor. If the stator field is weakened, the CEMF produced by the spinning armature is also reduced. This decrease in CEMF allows more current to flow in the rotor. Even with a weakened field, the increased current causes a higher speed. This higher speed comes at the expense of lower torque. At some point, the weakened field will not react with the high current armature to produce enough torque to keep the motor spinning. There is a sudden decrease in speed because the breakdown torque point has been reached.

ELECTRICAL BRAKING

Braking of the DC motor can also be accomplished by the controller functions. Two styles of braking can be used: regenerative braking and dynamic braking.

Regenerative braking is a form of electrical braking that provides a reversing voltage to the motor while it is still attached to the spinning load (see Figure 6–17). This circuit applies a reverse voltage to the stator, which attempts to drive the motor in the opposite direction. As the rotor attempts to align with the reversed field, a countertorque is produced that brings the motor to a quick stop. As the motor approaches zero speed, the reversed field is removed to prevent the motor from actually reversing.

Dynamic braking is another form of electrical brakes. In this style of braking, a resistor is inserted in the circuit to act as an electrical load for the spinning armature as normal armature current is disconnected. The armature has inertia from the driven load and continues to spin.

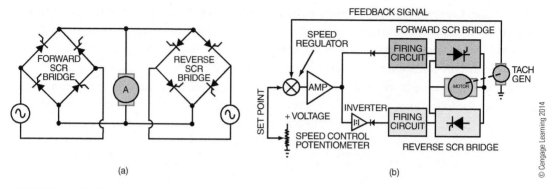

FIGURE 6–17 Regenerative braking.

The DC shunt field is maintained, and the armature now becomes a generator. If the armature is connected to an electrical load, the generator effects cause the rotor to slow with an electrical load attached. The resistor is called a dynamic braking resistor (DBR) and is a high-wattage, low-ohm resistor. It is usually located in a well-ventilated part of the controller so the dissipated heat is easily removed (see Figure 6–18).

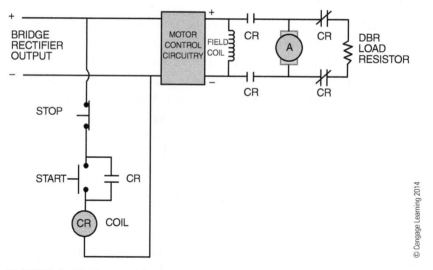

FIGURE 6–18 Dynamic braking circuit.

LOAD CONSIDERATIONS

Mechanical loads are characterized by the requirements of the driven machinery. The torque and horsepower requirements may change with the changing speed of the motor. These loads are classified as (1) constant horsepower, (2) constant torque, and (3) variable torque.

The following parameters are necessary to choose the right controller or to properly set up a DC drive for optimum control:

- Constant horsepower loads require more torque at low speed and less torque at high speed. The product of the torque times the speed must stay constant over varying speeds. Motor drives can be set to adjust the torque requirements based on the speed of the motor to produce a constant horsepower output.

- Constant torque loads have to maintain the same torque over a speed range. The horsepower of the motor varies. To overcome a friction load, as in conveyors and extruders, the torque is maintained over all speeds. More horsepower is required to start the load, but the horsepower can be reduced as the load moves up to speed.

- Variable torque loads are loads such as fans and pumps. In this type of load, the torque varies as the square of the speed. As the speed of a pump or fan increases, it does more work and higher torque is required.

SUMMARY

DC drives have many applications and many variables that must be considered when choosing the proper drive for the situation. Be sure you understand the needs of the driven machinery, to determine which drive or parameters are important. The drives all work from the same block-equipment approach. The AC input is converted to DC for the field. The AC is controlled by SCRs and a firing control block to rectify the AC and control the level of DC voltage and current sent to the armature. Various methods are used to provide safety, regulation, and speed setting. See Figure 6–19 for DC drive examples.

© Cengage Learning 2014

FIGURE 6–19 DC drive examples.

ACHIEVEMENT REVIEW

1. Name the concept used to reduce the speed of a DC motor below base speed.

   _____

   _____

2. What method can be used to increase speed above base speed?

   _____

   _____

   _____

3. Electronic shear pins are used to _____
   a. reduce the stress of the current on the motor windings.
   b. shut down the motor current if it rises too rapidly.
   c. hold the electronic firing of the SCRs to a shear level.
   d. break away under mechanical stress.

4. To increase the motor speed, the SCRs are allowed to conduct (more, less). (Circle the correct answer.)

5. Name a disadvantage of a half-wave rectifier system for the motor power.

   _____

   _____

   _____

6. The flywheel diode prevents _____
   a. inductive kick damage to the motor and drive.
   b. overvoltage conditions from damaging the motor during normal operation.
   c. cogging of the motor during slow-speed operations.
   d. the negative effects associated with motors that drive flywheel devices.

7. Open-loop control refers to the _____
   a. problem of the controller overshooting the design speed.
   b. method of measuring the actual current to the rotor.
   c. control system that does not provide speed information back to the controller.
   d. control system that provides speed information back to the controller.

8. Regulation of a DC motor speed means that the speed _____
   a. is held constant with various input frequencies.
   b. changes in proportion to the load.
   c. changes inversely to the load.
   d. stays constant under a varying load.

9. Regenerative braking requires a _____
   a. DBR.
   b. reverse field control.
   c. generator attachment to the motor.
   d. load that is able to be stopped immediately.

10. Variable torque loads _____
    a. change horsepower, but not torque, over a speed range.
    b. change speed, but not horsepower, over the range of the load.
    c. never change speed over the horsepower range of the drive.
    d. change torque exponentially as the speed changes.

UNIT 7

DC VARIABLE SPEED MOTOR CONTROL

BJECTIVES

After studying this unit, the student should be able to

- explain the operation of the CEMF method of acceleration for a DC motor.
- use elementary wiring diagrams, panel wiring diagrams, and external wiring diagrams.
- explain the ratings of starting and running protection devices.
- describe the operating principles of DC variable speed motor drives.
- state how above and below DC motor speeds can be obtained.
- list the advantages of DC variable speed motor drives.
- describe how solid-state devices can replace rheostats.
- make simple drawings of DC motor drives.
- list the advantages of using thyristors.

Although manual starters are still used, most industrial applications use automatic motor control equipment to minimize the possibility of errors in human judgment. To install and maintain automatic motor control equipment, the electrician must be familiar with three kinds of electrical circuit diagrams:

- schematic wiring diagrams
- panel wiring diagrams
- external wiring diagrams

The *electrical schematic wiring* diagram uses symbols and a simple plan of connections to illustrate the scheme of control and the sequence of operations.

The *panel wiring* diagram shows the electrical connections throughout all parts of the controller panel and indicates the external connections. All control elements are represented by symbols but are located in the same relative positions on the wiring diagram that they actually occupy on the control panel. Because of the maze of wires shown on the panel wiring diagram, it is difficult to use for troubleshooting or for understanding the operation of the controller. For this reason, the electrical schematic wiring diagram presents the sequence of operations of the controller, and the panel diagram is used to locate problems and failures in the operation of the controller.

The *external wiring* diagram shows the wiring from the control panel to the motor and to the pushbutton stations. This diagram is most useful to the worker who installs the conduit and the wires between the starter panel and the control panel and motor.

CEMF METHOD OF MOTOR ACCELERATION CONTROL

The CEMF produced by the rotor is low at the instant a motor starts. As the motor accelerates, this CEMF increases. Refer to voltage differential in the "Starting Current and CEMF" section in Unit 1. The voltage across the motor armature can be used to activate relays that reduce the starting resistance when the proper motor speed is reached.

Starting and Running Protection for a CEMF Controller

Starting protection for a CEMF controller is provided by fuses in the motor feeder and the branch-circuit line of the motor circuit. These fuses are rated according to *Article 430* of the *National Electrical Code®*.

Running protection for a controller, as defined in *Part III* of *NEC® Article 430*, is provided by an overload thermal element connected in series with the armature. The thermal element is rated at 115% to 125% of the full-load armature current. As covered in *NEC®* 430.32, if the current exceeds the percent of the rated armature current value, heat produced in the thermal element causes the bimetallic strip to open or trip the thermal contacts that are connected in the control circuit. The value of current during the motor start-up period does not last long enough to heat the thermal element sufficiently to cause it to open.

MOTOR CONTROL CIRCUIT

Starting the DC Motor

Refer to Figure 7–1. Close the main line switch before pressing the start button. After the start button is pressed, control relay M becomes energized. The control circuit is now complete from L_1 through the thermal overload (OL) contacts 6–7, through the start button contacts 7–8, and through the normally closed stop button contacts 8–9 to L_2. The lower auxiliary sealing contacts 7–8 of relay M also close and bypass the start button. As a result, the start button may be released without disturbing the operation.

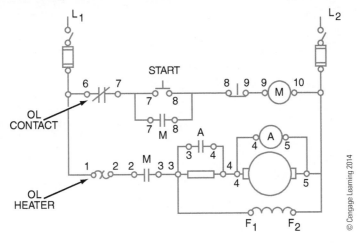

FIGURE 7–1 Elementary diagram of a DC CEMF controller.

When the main contacts 2–3 of contactor M are closed, the motor armature circuit is complete from L_1 through overload thermal element contacts 1–2, through contacts 2–3 of relay M, through the starting resistor, and through armature leads 4–5 to L_2. The shunt field circuit F_1–F_2 is connected in parallel with the armature circuit. Contacts 3–4 of CEMF contactor A remain open at start-up because a high inrush current establishes a high voltage drop across resistor 3–4. This leaves only a small voltage drop across the armature and the A contactor coil until acceleration is achieved. As the CEMF builds across the armature, it acts like a resistor that drops a larger percentage of the applied voltage, thus providing a voltage drop for the A coil and the A contacts 3–4 close to bypass current around the starting resistor.

Connecting the Motor across the Line

The CEMF generated in the armature is directly proportional to the speed of the motor. As the motor accelerates, the speed approaches the normal full speed, and the CEMF increases to a maximum value. Relay A is calibrated to operate at approximately 80% of the rated voltage. When contacts 3–4 of relay A close, the starting resistance 3–4 is bypassed, and the armature is connected across the line.

Running Overload Protection

A thermal overload relay contains two circuits. One circuit is in series with the armature and has the armature current flowing through its thermal sensor or heating element. The second circuit of the overload relay is the control circuit with a control contact. If the contact opens, because of excessive heat in the thermal heater, the control circuit will be interrupted and stop the motor. A thermal overload relay unit is shown in Figure 7–2. The schematic diagram is shown in Figure 7–3.

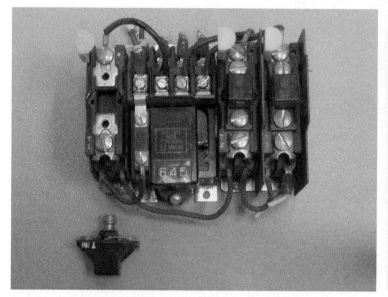
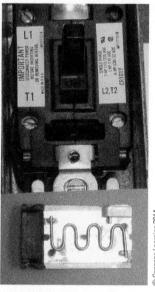

© Cengage Learning 2014

FIGURE 7-2 (A) A melting alloy–type overload relay with the left heater removed. (B) A bimetal-type overload assembly with the heater removed (under the bimetal relay).

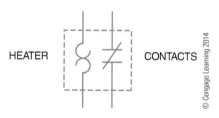

HEATER   CONTACTS

© Cengage Learning 2014

FIGURE 7-3 Schematic symbols for a thermal overload relay. The heater and contacts together make up the overload relay.

Refer to Figure 7–1 for the following explanation. When the load current of the armature exceeds the rated allowable percent of the full-load current, the overload thermal element (points 1–2) heats up and opens contacts 6–7 in the control circuit. Control relay M is de-energized, and main contacts 2–3 of M open and disconnect the motor from the line.

Stopping the Motor

When the stop button is pressed, the control circuit is broken at points 8–9. The same shutdown sequence occurs as in the case of the overload condition discussed previously. The sealing circuit 7–8 is broken in each case.

The advantage of this type of automatic starter is that it does not supply full voltage across the armature until the speed of the motor is correct. The starter eliminates human error that may result from the use of a manual starter.

PANEL WIRING DIAGRAM

Figure 7–4 shows the same CEMF control circuit presented in Figure 7–1. However, the panel wiring diagram locates the wiring on the panel in relationship to the actual location of the equipment terminals on the rear of the control panel. Troubleshooting or checking original installations requires an accurate comparison of the elementary schematic diagrams and panel diagrams. The electrician should use a system of checking connections on the diagram with the actual panel connections. For example, a colored pencil may be used to make check marks on the diagram as each connection is properly traced on the panel and compared to the diagram.

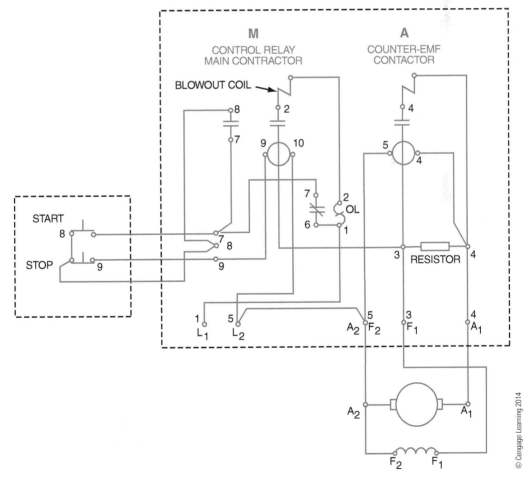

© Cengage Learning 2014

FIGURE 7-4 Panel wiring diagram for a DC CEMF motor controller.

CONDUIT OR EXTERNAL WIRING PLAN

All necessary external wiring between isolated panels and equipment is shown in the conduit plan (Figure 7–5). The proper size of conduit, size and number of wires, and destination of each wire are indicated on this plan. An electrician refers to this plan when completing the actual installation of the CEMF controller.

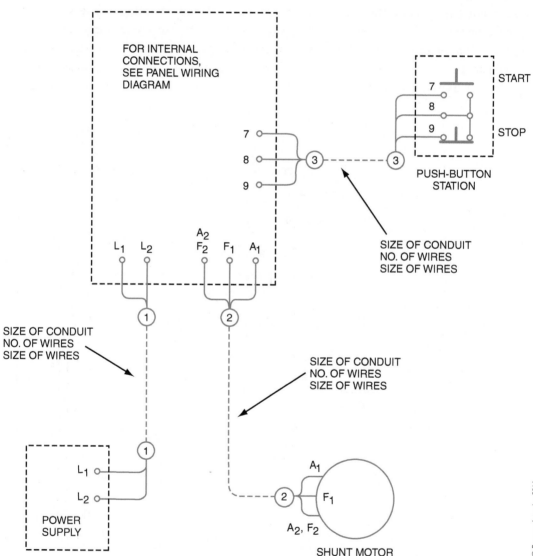

FIGURE 7–5 Conduit or external wiring plan for a CEMF controller.

© Cengage Learning 2014

DC ADJUSTABLE SPEED DRIVES

DC adjustable speed drives are available in convenient units that include all necessary control and power circuits.

Some machinery requirements are so precise that some AC variable frequency drives may not be suitable (see Unit 15). In such cases, DC motors provide characteristics that are not available on AC motors. A DC motor with adjustable voltage control is very versatile and can be adapted to a large variety of applications.

In the larger horsepower range, the motor-generator set used to be one of the most widely used methods of obtaining variable-speed control. The set consists of an AC motor driving a DC generator to supply power to a DC motor. Such motor-generator set drives, sometimes called *Ward-Leonard systems,* were used in early DC motor control and continued to be installed until the 1980s, to control the speed of the motor by adjusting the power supplied to the field of the generator, and, as a result, the output voltage to the motor (Figure 7–6). The generator field current can be varied with rheostats, as shown, or by variable transformers supplying a DC rectifier, or automatically with the use of solid-state controls. When it is desirable to control the motor field as well, similar means are used. Many of these systems are still in operation, but one is rarely installed as a new system.

The speed and torque of the system shown in Figure 7–6 can be controlled by adjusting the voltage to the field, or to the armature, or both. Speeds *above* the motor base speed (nameplate speed) are obtained by weakening the motor shunt field. Speeds *below* the motor base speed are obtained by weakening the generator field. As a result, there is a decrease in the generator voltage supplying the DC motor armature. The motor should have a full shunt field for speeds lower than the base speed, to give the effect of continuous control, rather than step control of the motor speed.

The motor used to furnish the driving power may be a three-phase induction motor, as shown in Figure 7–6. After the driving motor is started, it runs continuously at a constant speed to drive the DC generator.

The armature of the generator is coupled electrically to the motor armature as shown. If the field strength of the generator is varied, the voltage from the DC generator can be controlled to send any amount of current to the DC motor. As a result, the DC motor can be made to turn at many different speeds. Because of the inductance of the DC fields and the time required by the generator to build up voltage, extremely smooth acceleration is obtained from zero r/min to speeds greater than the base speed.

The field of the DC generator can be reversed automatically or manually, with a resulting reversal of the motor rotation.

The generator field resistance can be changed automatically by using SCRs (or thyristors) or time-delay relays operated by a CEMF across the motor armature. The generator field resistance can also be changed manually.

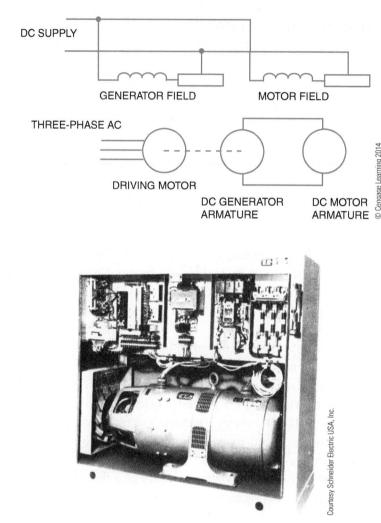

FIGURE 7-6 (A) Basic electrical theory of a DC motor–generator, variable speed control system. (B) Packaged motor generator with DC variable speed control system supplied from AC.

Electrically controlled, variable speed motor drives offer a wide choice of speed ranges, torque, and horsepower characteristics. They provide a means for controlling acceleration and deceleration, and methods of automatic or manual operation. A controlling tachometer feedback signal may be driven by the DC motor shaft. This is a system refinement to obtain a preset constant speed. This method depends on the type of application, speed, and degree of response desired. In addition to speed, the controlling feedback signal may be set to respond to pressure, tension, shock, or some other transducer function.

One of the most advantageous characteristics of the motor-generator set drive is its inherent capability to regenerate. In other words, when a high-inertia load overdrives the motor, the

DC motor becomes a generator and delivers reverse power. For example, assume the DC motor is running at base speed. If the generator voltage is decreased by adjusting the rheostat to slow the motor, the motor counter-voltage will be higher than the generator voltage, and the current reverses. This action results in reverse torque in the motor, and the motor slows down. This process is called dynamic braking. This dynamic feature is very desirable when used on hoists for lowering heavy loads, metalworking machines, textile and paper processing machines, and in general industry for the controlled stopping of high-inertia loads. Multiple motor drives are also accomplished with this type of motor-generator drive.

Motor-generator set drives using automatic regulators have been used for years for nearly every type of application. A higher degree of sophistication in controls has been developed, making it possible to meet almost any desired level of precision or response.

STATIC MOTOR CONTROL DRIVES

Despite the previous use of the motor-generator drives, rotating machines were required to convert AC to mechanical power. As a result, the combined efficiency of the set is rather low; it requires the usual rotating machine maintenance, and it is noisy. *Static* DC drives now being used have no moving parts in the power conversion equipment that converts (rectifies) and controls the AC power (Figures 7–7 and 7–8). The solid-state devices are used for controlled conversion of AC line power to DC.

The basic theory for obtaining DC motor speeds below and above base speed is the same as with a motor-generator set. Only the method of controlling the voltages and field strengths differs. For example, in Figure 7–9, the armature is supplied with DC rectified from an AC source. The AC is rectified by the use of the SCR in the controlled circuit to obtain DC. The gate of the thyristor turns on the SCR at the proper portion of the half-wave, thereby controlling the motor below base speed. Figure 7–9 is a simplified circuit for the purpose of illustration. The field strength would be held at its fullest strength in a similar manner. For speeds above the motor base speed, the field control can weaken field strength with full armature voltage. The feedback tachometer maintains a preset speed.

In Figure 7–9, the SCR is controlled by the setting of the potentiometer, or *speed control.*

FIGURE 7–7 Control panel for SCR-controlled DC motor drive.

© Cengage Learning 2014

FIGURE 7–8 SCRs in various sizes.

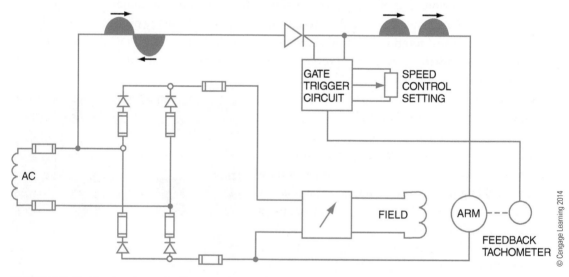

FIGURE 7–9 Single-phase, half-wave armature controlling a small motor.

This varies the "on" time of the thyristor per AC cycle, and thus varies the amount of average current flow to the armature. When speed control above the base speed is required, the rectifier circuit in the field is controlled by SCRs, rather than diodes.

The SCR, or thyristor, can control all the positive waveform or voltage through the use of a method called *phase shifting*. Covering the theory of the method is beyond the scope of this book.

The SCR is probably the most popular solid-state device for controlling large and small electrical power loads. The SCR is a controlled rectifier that controls an electric current. It will not conduct when the voltage across it is in the reverse direction. It will conduct only in the forward direction when the proper signal (voltage) is applied to the gate terminal (see Figure 7–10). The gate is normally controlled by electronic pulses from a control circuit.

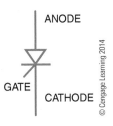

ANODE

GATE   CATHODE

FIGURE 7–10 Schematic symbol for an SCR.

The gate will turn the SCR on but will not turn it off in a DC circuit. To turn the anode-cathode section of the SCR on (close the switch), the gate must be the same polarity as the anode with respect to the cathode. After the gate has turned the SCR on, it remains on until the current flowing through the power circuit (anode–cathode section) is either interrupted or drops to a low enough level to permit the device to turn off. The anode-to-cathode current must fall below the holding current level. The *holding current,* or *maintaining current,* is the amount of current required to keep the SCR turned on. The SCR performs the same function as a rheostat would in controlling motor field strengths or voltage to an armature. It is similar to a variable resistance because it can be adjusted throughout its power range. The SCR control has replaced the rheostat because it is smaller in size for the same current rating, more energy efficient, and cheaper.

SUMMARY

DC motors need controls to start, stop, protect, and adjust the speed and torque of the motor. The systems used must comply with the *NEC®* and also have approval from testing firms such as Underwriters Laboratories, Inc. The motors must be protected from overheating and causing damage to the motor and the surrounding area. This unit introduced the two general styles of wiring diagrams: the schematic, which shows the electrical operating sequence of the components, and the wiring diagram, which shows the physical relationship of the equipment. Motor generator sets were presented to familiarize you with the possible sources of DC control. At present, much of the control is done by solid-state DC electronic drives.

ACHIEVEMENT REVIEW

Select the correct answer for each of the following statements, and place the corresponding letter in the space provided.

1. The least important plan or diagram in troubleshooting motor controls is probably the

    _____

    a. electrical schematic plan.
    b. panel diagram.
    c. external conduit plan.
    d. layout of the area in which the controllers are installed.

2. The best diagram to use for determining how a controller operates is the _____
    a. electrical schematic plan.
    b. panel plan.
    c. external plan.
    d. architectural plan.

3. The physical location of control wires is shown on the _____
    a. electrical schematic plan.
    b. architectural plan.
    c. conduit plan.
    d. panel wiring diagram.

4. The DC CEMF controller results from the automatic actions of the _____
    a. applied voltage.
    b. changing voltage across the armature.
    c. changing voltage across the field.
    d. starting current.

5. Overload protection is the same as _____
    a. starting protection.
    b. mechanical protection.
    c. electrical protection.
    d. running protection.

6. Overload contacts open the circuit when the motor current reaches _____
    a. 85% of full load.
    b. 100% of full load.
    c. 125% of full load.
    d. 150% of full load.

7. In the event a motor is allowed to exceed the permissible value, it is protected by

    _____
    a. starting protection.
    b. fuses.
    c. an overload thermal element.
    d. the stop button.

8. With the disconnect switch closed, the shunt field in Figure 7–1 is placed across the line when the _____
   a. A contact closes.
   b. disconnect switch is closed.
   c. M contact closes.
   d. start button closes.

9. In Figure 7–1, contact A is closed when the _____
   a. start button is closed.
   b. stop button is opened.
   c. A coil is de-energized.
   d. A coil is energized.

10. The motor in Figure 7–1 is placed across the line when _____
    a. the start button is closed.
    b. the disconnect switch is closed.
    c. contact A is closed.
    d. contact M is closed.

11. What is the DC motor base speed? _____

    _____

    _____

    _____

12. How is the speed of a DC motor controlled above the base speed? _____

    _____

    _____

    _____

13. How is the speed of a DC motor controlled below the base speed? _____

    _____

    _____

    _____

14. How may an SCR replace a rheostat? _____

_____

_____

_____

15. List the advantages of using thyristors in the motor drive control. _____

_____

_____

_____

UNIT

DC MOTOR
DYNAMIC BRAKING

8

OBJECTIVES

After studying this unit, the student should be able to

- list the steps in the operation of a DC motor control with interlocked forward and reverse pushbuttons.

- explain the principle of dynamic braking.

- describe the operation of a CEMF motor controller with dynamic braking.

Industrial motor installations often require that motors be stopped quickly and that the direction of rotation be reversed immediately after stopping. To achieve this operation, electrically and mechanically interlocked pushbutton stations connected to relays are used to disconnect the armature from the supply source. The armature is then connected to a low value of resistance. Because the inertia of the armature and connected load causes the armature to continue to revolve, it acts as a loaded generator. As a result, the armature is slowed in speed. This action is called *dynamic braking*.

Reversal of motors and dynamic braking are operations used in special equipment such as cranes, hoists, railway cars, and elevators.

MOTOR-REVERSAL CONTROL

A DC motor is reversed by reversing the armature connections. The type of compounding is not affected by this method of obtaining reversal.

The pushbutton control station illustrated in Figure 8–1 is the type used for motor reversal. The forward and reverse buttons are mechanically interlocked so that it is not possible to operate these buttons at the same time.

FIGURE 8-1 A forward, reverse, stop pushbutton station.

© Cengage Learning 2014

Description of Operation

Forward Starting. When the forward button is pressed, the normally open, forward momentary contacts close and the normally closed, forward momentary contacts open. The control circuit is shown in Figure 8–2. The forward contactor coil is energized from L_1 through the overload contacts, stop button, forward pushbutton contacts 1–2 (when closed), and reverse button contacts 3–4 through the forward contactor coil to L_2. The forward (F) contacts are a holding circuit in the forward pushbutton. In the power circuit (Figure 8–2), the F contacts of the forward energized contactor close and thus complete the armature circuit through the starting resistance. The normal CEMF starter sequence of operations then continues to completion.

Reverse Operation. If the reverse pushbutton is pressed, contacts 3–4 of the reverse button open and thus de-energize the forward contactor coil F. In addition, the F contacts are opened as well as the sealing F contacts. Pressing the reverse button also completes the circuit of the reverse contactor coil (R) that closes the R contacts. The motor armature circuit is now complete from L_1 to A_2 and A_1 to L_2 (Figure 8–2). The armature connections are reversed, and the armature rotates

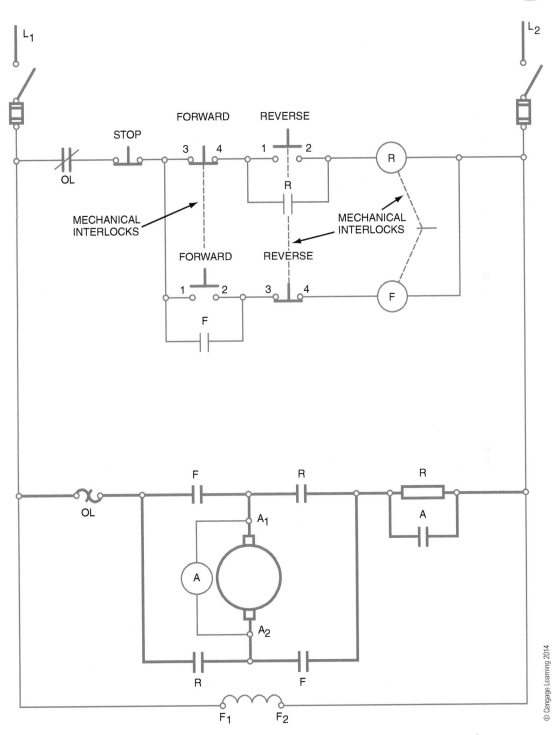

FIGURE 8–2 Electrically and mechanically interlocked control and powered circuit for reversing motor.

in the opposite direction. It is impossible for the reverse contacts to close until the forward contacts are open, because of the electrical and mechanical interlocking system used in this type of control circuit. The mechanical interlocks are shown by the broken lines between the R and F coils in Figure 8–2.

Dynamic Braking

The purpose of dynamic braking is to bring a motor to a quicker stop. To do this, there must be a method to quickly use the mechanical energy stored in the momentum of the spinning armature after the main switch is opened. One method is to change the function of the motor to that of a generator. (A generator converts mechanical energy into electrical energy.) At the instant the motor is disconnected from the line, a resistor is connected across the motor armature. The resistor loads the motor as a generator, dissipates the electrical and therefore the mechanical energy, and slows the motor quickly.

DYNAMIC BRAKING USED IN A CEMF CONTROLLER

As an example, the principle of dynamic braking is shown by following the steps in the operation of an elementary CEMF controller (Figure 8–3).

This analysis emphasizes the dynamic braking operation rather than the details of the circuit, which were presented previously. The dynamic braking magnetic (DBM) coil is designed so that its only function is to ensure a positive closing of the normally closed dynamic braking contacts 9–10. If the main coil M is energized, the dynamic braking contacts 9–10 open from the Normally closed condition, and contacts 8–9 of M are closed, although the dynamic braking coil is also energized.

When the start button is pressed, the control coil M is energized, contacts 8–9 of M close, and the motor starts and accelerates up to normal speed by the CEMF method. At the instant the M control relay is energized, the DBM coil energizes and the normally closed, dynamic braking contacts 9–10 open. As a result, the dynamic brake resistor (DBR) connection across the armature is broken.

Stopping

When the stop button is pressed, relay control coil M is de-energized, and M contacts 8–9 open the armature circuit, de-energize the DBM coil, and close the dynamic braking contacts 9–10. These contacts connect the dynamic braking resistor directly across the armature. Because the shunt field is still connected across the line and receiving full excitation, the high CEMF generated in the armature causes a high load current through the dynamic brake resistor. The heavy load current dissipates the stored mechanical energy in the armature acting as a heavily loaded generator, with the result that the motor slows to a stop. Braking action decreases as the armature speed decreases.

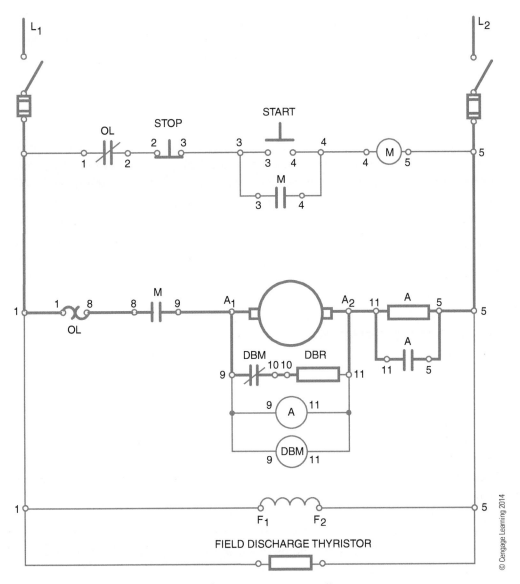

FIGURE 8-3 Elementary diagram for a DC CEMF controller.

Field Discharge Resistor

When using the CEMF controller *and* dynamic braking, a field discharge resistor must be added. When the shunt field is disconnected from the supply voltage, its magnetic field begins to collapse. The quick collapse of the magnetic field produces a very large "inductive kick" voltage—possibly thousands of volts. If it does not have a discharge path, the high voltage actually begins to break down the field winding insulation. Normally the discharge path is through the armature, which allows a slower collapse and keeps the voltage small. A field discharge resistor is a thyristor-type device that conducts when the voltage across it is high enough, but has a high resistance to normal line voltage. The collapsing magnetic field's voltage can discharge through the field discharge resistor (FDR) without damaging the field windage.

Summary

Figure 8–2 shows the reversing circuit with electrical pushbutton interlocks and mechanical interlocks between the forward and reverse contactor. The motor is reversed by reversing current flow through the armature but keeping the shunt field current in the same direction. Dynamic braking is used when slowing the motor. As the armature is disconnected from the power source, a dynamic braking resistor is connected across the armature. The armature is still spinning and the shunt field is still energized, so the armature acts as a generator. The resistor provides a current path for the generated current and slows the armature as it works as a generator. A field discharge resistor is also used to prevent the sudden collapse of the shunt field flux as it is disconnected from the line power.

Achievement Review

1. How can DC motors be reversed without changing the type of compounding?

   _____

   _____

   _____

2. What interlocking is necessary on the forward and reverse pushbuttons to avoid short circuits? _____

   _____

3. What will happen if the forward and reverse relays are energized at the same time?

   _____

   _____

   _____

   _____

4. How many contacts are required on the forward relay? _____

5. What are two applications requiring motor reversal? _____

_____

_____

6. How can a motor be shut down quickly without using a mechanical brake?

_____

_____

_____

7. When is dynamic braking applied? _____

_____

_____

_____

8. How does dynamic braking slow a motor?_____

_____

_____

_____

9. If an electromagnetically operated brake is connected in series with the armature, how is
    it operated? _____

_____

_____

_____

_____

10. Name two installations in which dynamic braking is used._____

_____

_____

UNIT 9

BASIC PRINCIPLES OF AUTOMATIC MOTOR CONTROL

OBJECTIVES

After studying this unit, the student should be able to

- list several factors to be considered when selecting and installing electric motor control equipment.
- explain the purpose of a contactor.
- describe the basic operation of a contactor and relay.
- list the steps in the operation of a control circuit using start and stop pushbuttons.
- interpret simple automatic control diagrams.
- draw a simple magnetic control circuit.

Motor control was a simple problem when motors were used to drive a common line shaft to which several machines were connected. In this arrangement, it was necessary to start and stop only a few times daily.

With individual drives, however, the motor is an integral part of the machine, and the motor controller must be designed to meet the needs of the machine to which it is connected.

As a result, the modern motor controller may not just start, stop, and control the speed of a motor. The controller may also be required to sense a number of conditions, including changes in temperature, open circuits, current limitations, overload, smoke density, level of liquids, or the position of devices. Manual control is limited to pressing a button to start or stop the entire sequence of operations at the machine or from a remote location.

The electrician must know the symbols and terms used in automatic control diagrams to be able to wire, install, troubleshoot, and maintain automatic control equipment.

CLASSIFICATION OF AUTOMATIC CONTROLLERS

Purpose

Factors to be considered when selecting motor controllers include the required types of starting, stopping, reversing, running, and controlling speed and sequence. Other factors that influence the selection of a controller include the electrical service, environmental conditions, and electrical codes and standards.

Operation

The motor may be controlled directly or manually by an operator using a switch or a drum controller. Remote control uses contactors, relays, pushbuttons, sensors, and possibly electronics.

CONTACTORS

Contactors, or relays (Figure 9–1 and Figure 9–2), are required in automatic controls to transmit varying conditions in one circuit in order to influence the operation of other devices in the same or another electrical circuit. Relays have been designed to respond to one or more of the following conditions:

voltage	overvoltage	undervoltage
current	overcurrent	undercurrent
current direction	differential current	power (watts)
power direction	volt-amperes	frequency
phase angle	power factor	phase rotation
phase failure	impedance	speed
	temperature	

FIGURE 9–1 DC magnetic relay.

FIGURE 9–2 DC operated relay.

Magnetic switches are widely used in controllers because they can be used with remote control and are economical and safe.

A relay or contactor usually has a coil that can be energized to close or open contacts in an electrical circuit. The coil and contacts of a relay are represented by symbols on the circuit diagram or schematic of a controller. Symbols commonly used to represent contactor elements are shown in Figure 9–3.

If the control coil is connected in series in the motor power circuit, the heavy line symbol shown at the left of Figure 9–3 is used. If the coil is connected in parallel (shunt), the light line symbol is used.

A series coil has a large current-carrying conductor with few turns designed to carry large currents. A shunt coil has a small wire size with many turns; it carries small currents. It is possible for a series coil and a shunt coil to have the same ampere turns, producing similar magnetic results.

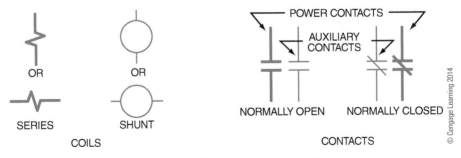

FIGURE 9–3 Schematic symbols for contactor elements.

Contacts that are open when the coil is de-energized are known as *normally open* contacts and are indicated by two short parallel lines. Contacts that are closed when the coil is de-energized are called *normally closed* contacts and are indicated by a slanted line drawn across the parallel lines.

To minimize heavy arcing that burns the contacts, a DC contactor usually is equipped with a *blowout coil* and an *arc chute*. Figure 9–4 shows a magnetic contactor that is provided with a blowout coil and an arc chute.

As the contacts move apart, an arc is established between the contact points. The arc is still conducting current and acts like a conductor. If a conductor is placed in a magnetic field, it tries to move out of the magnetic field (motor action). Figure 9–5(A) shows that the current is still flowing through the magnetic blowout coil to the arc. This current in the blowout coil creates a magnetic field as long as there is current in the arc. The magnetic field deflects the arc away from the contact surfaces, elongates the arc, and "blows" it into the arc chute to be extinguished. As soon as the arc is stretched far enough and broken into small segments, the current flow is interrupted and the blowout coil has no more current and no magnetic effect. Figure 9–5(B) shows a contact without blowout coils. In this case, the arc is allowed to stretch and rise through thermal action. As it rises, it is elongated and finally breaks apart at the tip of the arc horns rather than burning the contact closure surface.

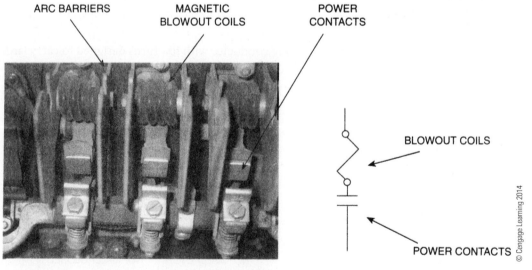

FIGURE 9–4 Magnetic blowout coils magnetically move the arc away from the contacts.

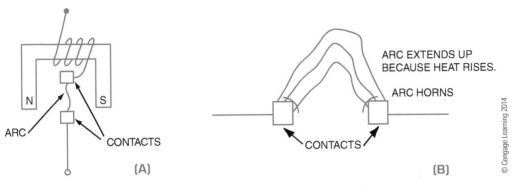

FIGURE 9–5 (A) Magnetic blowout coils are used for arc control on large power contacts. (B) Contacts without blowout coils rely on the arc stretching to break.

PUSHBUTTONS

Pushbutton stations (Figure 9–6) are spring-controlled switches, which, when pushed, are used to complete motor or motor control circuits. Figure 9–6(B) shows multiple control stations with pushbuttons, selector switches, and pilot indicating lights. Note the "mushroom" stop button for easy emergency access. This is often referred to as an *E stop,* short for "emergency stop."

(A)

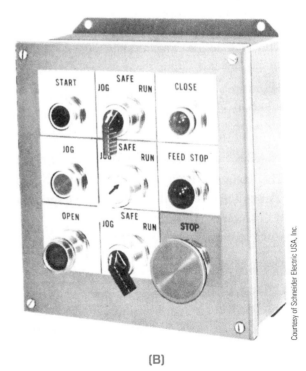

(B)

FIGURE 9–6 Pushbutton stations.

The symbols used in schematic drawings to represent momentary pushbutton contacts are provided in Figure 9–7. Contacts can either be normally open or normally closed. This is the normal condition when there is no mechanical actuation of the contacts. In the pushbuttons shown in Figure 9–7, the contacts are referred to as *momentary contacts.* This simply means that the contacts change from their *normal* condition to the opposite condition momentarily when mechanical actuation is applied, and then change back to the normal condition when the actuator is removed. Some contacts are designated as *maintained* contacts. This means that the contacts stay as activated (held mechanically) until returned to their original position.

NORMALLY OPEN   NORMALLY CLOSED   OPEN AND CLOSED

© Cengage Learning 2014

FIGURE 9–7 Symbols for pushbutton contacts.

TYPICAL CONTROL CIRCUIT

Figure 9–8 shows an elementary control circuit with momentary start and stop buttons and a sealing, or "holding," circuit. The typical control circuit uses an electromagnetic coil to move sets of contacts. The contacts move to open and close the power circuit to the motor and also to open and close contacts in the control circuit. The control contacts provide a holding circuit in parallel to the start momentary contacts. This parallel circuit is referred to as the *sealing circuit* or *holding circuit.* It seals a current path around the normally open start button contacts. The circuit operation is as follows:

1. When the start button is pressed to close contacts 2–3, current flows from L_1 through normally closed contacts 1–2 of the stop button, through the closed contacts 2–3 of the start button, and through coil M to line L_2.

2. The current in coil M causes the contact M to close. Thus, the holding circuit around contacts 2–3 of the start button closes. The start button may now be released, and even though the spring of the pushbutton opens contacts 2–3, coil

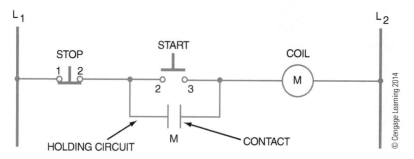

FIGURE 9–8 A control circuit with start and stop buttons and holding circuit.

M remains energized and holds contacts M closed to maintain a holding circuit around the normally open contacts 2–3 of the start button. Coil M, being energized, also closes M contacts in the power circuit to the motor (not shown).

3. If the stop button is momentarily pressed, the circuit is interrupted at contacts 1–2, and coil M is de-energized. Contacts M then open, and coil M cannot be re-energized until the start button again closes contacts 2–3.

SCADA AND HMI

SCADA, or supervisory control and data acquisition, also sometimes known as supervisory calculation and data analysis, is a system that uses input and output data from real-world processes and machinery interfaces. These data that are sent to the processors, usually through a programmable logic controller (PLC), are used to monitor, control, and affect the output of the PLC to determine control operations. The PLC operates to control the output functions with predetermined programs and therefore monitors and controls the assigned process. In order for humans to interact with the processor, some method of communication is needed. Typically this interface to the PLC started out as a program panel, but more recently a personal computer has been used. The personal computer is still used to design and develop the operating program, but the monitoring and changes to the operations can be processed through the human machine interface (HMI). The HMI, as seen in Figure 9–9, is a real-time

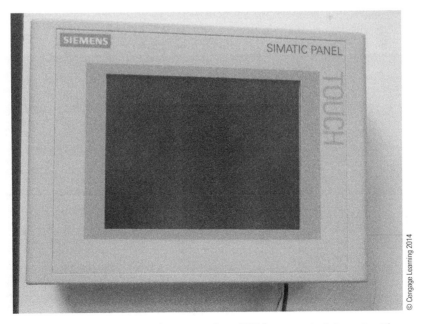

FIGURE 9–9 Human machine interface (HMI) screen to interact with control systems.

interface to the data acquisition system and a control point so the human operators can change operations based on their decisions.

Often the data are acquired over networks to provide the inputs to the SCADA system. If terminals are used for the interface, they are often referred to as remote terminal units (RTUs). In most cases, SCADA systems run without human intervention as automatic controls or monitoring systems for automatic systems, and the RTU is used for reference and control only if needed. In the event that more human control is desired or needed, the current trend is to use HMIs as interfaces to the control system. The HMI can be connected to the RTU or the PLC to take data and display them in different format.

An HMI can be a control panel with hard button interfaces to change operations of the PLC and therefore the process. The buttons may have multiple functions depending on which program the human operator chooses. The current trend is to make the control buttons as virtual controls that are programmable and can respond to the touch screen of the HMI interface. In this case, the operations are displayed as graphical elements and the controls can be programmed to respond when people simply touch the screen at the point where they want to make a change or analyze the process. See Figure 9–10.

SCADA systems, RTUs, and HMIs have become more reliable and more dependable. Some automatic safety systems and emergency stop systems have been approved as SCADA systems without the need for redundant hardwired systems. The SCADA and HMIs are typically linked to a historical data system that allows the operators to see data trends and failure modes and track maintenance requirements as a result of equipment wear or sensor breakdown. The manufacturing and materials-handling industries can predict maintenance with great certainty and help with failure analysis based on the SCADA systems.

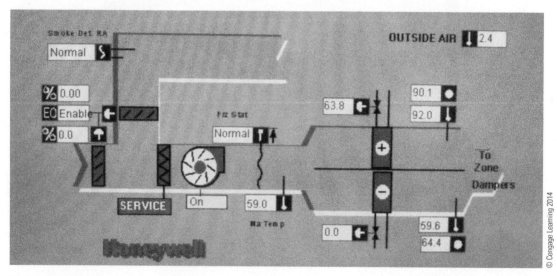

© Cengage Learning 2014

FIGURE 9–10 Sample display screen to show data collection and real-time controls.

SCADA connections and RTUs and HMIs are increasingly being connected through systems that use an open protocol. This means that each individual manufacturer no longer has a specific protocol base that allows only specific compatible hardware to be connected. This was done in the past to force end users to buy from a specific supplier if they wanted the systems to "talk" to each other. Current systems allow competitors to connect to each other's systems, thereby allowing the end users to pick the most effective system to meet their needs. The perceived drawback of open protocol may be that more people can tie into the networks and wreak havoc with critical processes either on purpose or by accident. The fact that major systems such as power plant control or water treatment facilities control could become vulnerable to malicious intent is of concern.

SUMMARY

The basic automatic control circuit is used to control larger motors through electromagnetic relays. This allows the operation to be remotely located and the contactor to be located near the motor. The basic principle uses a momentary-contact switch to close a circuit to a magnetically operated relay.

ACHIEVEMENT REVIEW

Select the correct answer for each of the following statements, and place the corresponding letter in the space provided.

1. Early motor installations consisted of _____
   a. individual drives.
   b. a common line shaft drive.
   c. automatically controlled motor drives.
   d. remotely controlled motors.

2. Individual motor drives require _____
   a. single-phase motors.
   b. automatic controllers.
   c. speed rheostats.
   d. gear heads.

3. Automatic DC motor controllers are designed to respond to in temperature, open circuits, current limitations, and _____
   a. wire size.
   b. fuse rating.
   c. speed acceleration.
   d. brush assembly.

4. Interpretation of automatic control circuits requires the recognition of _____
    a. color.
    b. electrical circuit symbols.
    c. ratings.
    d. parallel circuits.

5. A relay symbol shows the _____
    a. number of turns in a coil.
    b. relay current rating.
    c. relative position of the component parts.
    d. size of the contacts.

6. A relay is classified as a piece of electrical equipment at least one_____
    a. coil.
    b. resistor.
    c. coil operating one contact.
    d. coil operating two contacts.

7. Normally open contacts are_____
    a. open at all times.
    b. open when the relay coil is de-energized.
    c. open when the relay coil is energized.
    d. contacts that open a circuit.

8. Normally closed relay contacts are represented by the symbol:_____
    a.
    b.
    c.
    d.

© Cengage Learning 2014

9. A holding circuit bypasses_____
    a. the armature circuit.
    b. the field circuit.
    c. the "on" pushbutton contacts.
    d. the relay coil.

10. Schematic control diagrams are read from_____
    a. top to bottom.
    b. bottom to top.
    c. right to left.
    d. field to armature circuit.

UNIT

ELECTRO-MECHANICAL AND SOLID-STATE RELAYS AND TIMERS

BJECTIVES

After studying this unit, the student should be able to

- explain how relays operate.
- list the principal uses of relays.
- describe different relay control and load conditions.
- tell how SCRs operate.
- identify relay component symbols.
- connect different relays in a circuit.
- identify and use various timers.
- use proper timer symbols in schematic diagrams.

RELAYS

Relays are devices used to relay or multiply control signals or electrical contact closures. The relay concept is used where a small voltage at low current operates a set of electrical contacts to an open or closed position. This contact operation in turn controls a larger electric load as it relays the electrical operations.

Another common use of a relay is to multiply a single signal to open or close multiple contacts to control multiple electrical loads.

ELECTROMECHANICAL RELAYS

Electromechanical relays, contactors, and motor starters basically operate by the same principles. These electrically operated switches respond to the electromagnetic attraction of an energized coil of wire mounted on an iron core. The devices differ in the amount of current that each associated contact must switch. The relay—which can be compared to an amplifier—is usually used to switch small amounts of currents (usually 0–15 amperes) in many control circuits (Figure 10–1). Uses of relays include switching (on and off) larger coils of motor starters, contactors, solenoids, heating elements, and small motors. Other uses are alarm systems and pilot light control. Relays have many industrial and commercial applications, both AC and DC.

A small current flow and/or low voltage applied to a relay coil can result in a much larger current or voltage being switched. One input signal (voltage) may control several output (switched) circuits (Figure 10–2).

The coil voltage of the relay is separate or different from those at the switched contacts; this is called *separate control*. However, the coil voltage may be the same system voltage as the switched voltage.

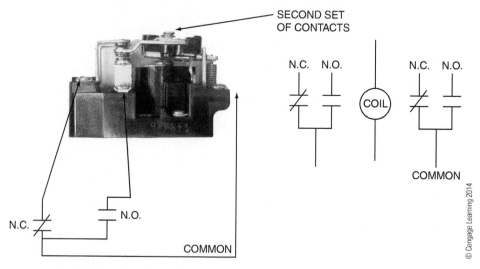

FIGURE 10–1 Control relay and associated coil contact diagram.

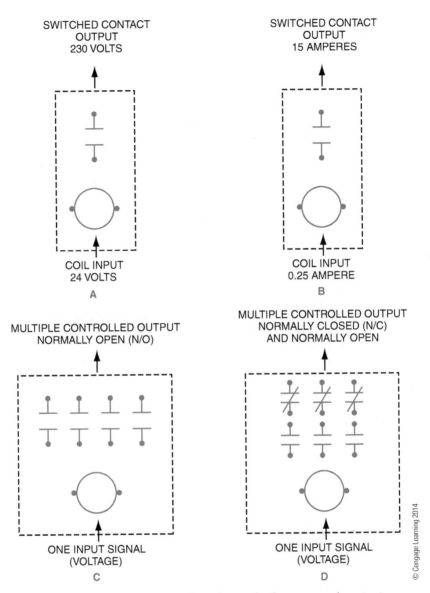

FIGURE 10-2 Several electrical-mechanical relay uses and contact configurations.

Relays are available in many shapes and sizes. Some are sealed in dustproof, transparent plastic enclosures, shown in Figure 10–3. The general construction of a typical relay is shown in Figure 10–4.

Note in Figure 10–2 that relay contacts may be normally closed or normally open. The action of the contacts is to switch something "on" or "off" depending on the configuration.

© Cengage Learning 2014

FIGURE 10-3 "Ice-cube" relays and plug-in bases for easy replacement.

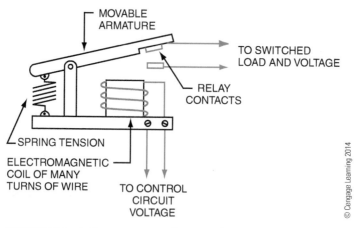

© Cengage Learning 2014

FIGURE 10-4 Construction of a typical electromagnetic control relay.

SOLID-STATE RELAY (SSR)

A solid-state relay (SSR) can be used to control most of the circuits that the electromechanical relay controls. By comparison, the SSR has no coil or contacts. The semiconductor industry has developed solid-state components with unusual applications to the industrial control processes. These components are compact, versatile, and very reliable if used in the proper application.

The silicon-controlled rectifier (SCR) is probably the most popular solid-state device for controlling large and small electrical power loads. Basically, the SCR either conducts or does not conduct an electric current. When it is not conducting, the SCR offers almost a complete blockage to the current. It passes only a few milliamperes to the load. For this reason, some manufacturers place contacts from electrically operated contactors in the total circuit to disconnect the load completely.

The SCR will not conduct when the voltage across it is in the reverse direction. It conducts only in the forward direction when the proper signal (voltage) is applied to the gate terminal (Figure 10–5). Once it is conducting, the SCR cannot be turned off immediately. It is necessary only to provide a small signal to start the SCR conducting a current. It continues to conduct even without a signal from that point on, as long as the current is in the forward direction. The only way to stop the SCR from conducting is to reduce the current flow below the holding current level or disconnect it from the system. On AC, of course, this happens every half cycle, so this characteristic is no problem (Figure 10–6). For DC applications, the voltage is reduced to zero by interrupting the circuit, generally with a contact on an electromagnetic relay.

Figure 10–7 shows a typical SSR. Note the input and switched output terminal connections. Figure 10–8 shows these connections completed. Terminal wiring is very simple and consists of two input control wires and two output load wires. The connecting terminals are clearly identified on SSRs (as they are on electromechanical relays). The relay in Figure 10–8 has a light-emitting diode (LED) connected to the input or control voltage. When the input voltage turns the LED on, a photodetector connected to the gate of the TRIAC turns the TRIAC on and connects the load to the line. This optical coupling is commonly used with SSRs. These relays are referred to as being *optoisolated*. This means that the load side of the relay is optically isolated from the control side of the relay. The control medium is a light beam. No voltage spikes or electrical noise produced on the load side of the relay is therefore transmitted to the control side.

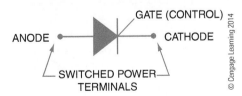

FIGURE 10–5 Schematic symbol for an SCR, which is the heart of the solid-state relay.

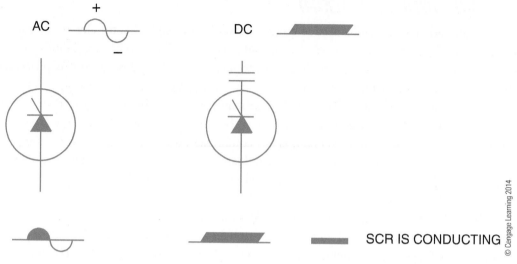

FIGURE 10–6 An SCR contacts current in the forward direction until the voltage is reduced to zero.

© Cengage Learning 2014

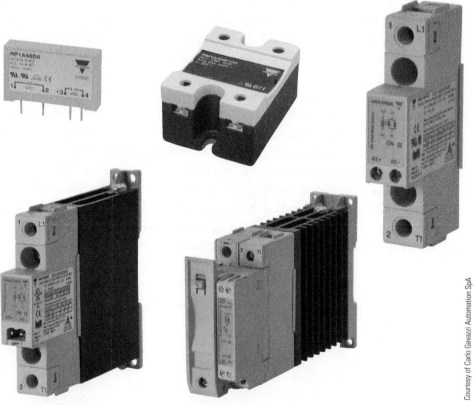

FIGURE 10–7 Typical package for SSRs.

Courtesy of Carlo Gavazzi Automation SpA

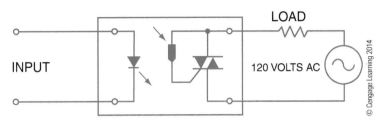

FIGURE 10-8 SSR used to control an AC load.

SSRs have a wide range of input or control voltage designs. The SSRs may operate at TTL (transistor–transistor logic) levels plus 5 volts DC. They may operate at levels between 3 and 30 volts DC or other relays at between 90 and 120 volts AC. Considerations as to what style of relay include the way that the SSRs turn on.

SSRs are available with a design to match the requirements of the application. An SSR with *zero switching* capability is used for resistive, inductive, and capacitive loads. Zero switching refers to when the SSR's output is activated. When the control signal voltage is applied, the SSR output turns on at the first zero crossing of the line voltage; therefore, the response time is typically less than a half cycle. These are the most commonly used SSRs.

Analog switching is used for resistive loads. The input control can be varied from 4 to 20 milliamps (mA). This SSR can then perform phase control when the output is controlled by adjusting the input voltage and controlling the "on" time of the load. These SSRs can be used to reduce inrush currents or to control heating elements for precise heat requirements.

Analog full-cycle switching is also used for resistive loads. This SSR responds to 4–20 mA or 1–10 volts DC control values. A low value of input control turns on the output and leaves it on for a specific number of output cycles. A high value of input turns on the output for a full "on" state of the load. These relays are used for analog control of heating elements with varying input values. *Instant on switching* is used for inductive loads. The SSR is immediately and fully "on" after application of the control voltage. The output turns on anywhere during the cycle, and the response time is typically less than 1 millisecond. This relay is used to control solenoids or coils where immediate response is needed. *Peak switching* is used for heavy industrial loads such as transformer loads. The output will turn on at the first AC peak value after application of the control voltage. The last switching state is a *DC switching SSR*. The application of a DC input turns on the DC voltage output. The relay is used for controlling DC loads. When this relay is used for coil loads, such as DC motors and coils, then an inductive-kick protection diode is needed to prevent damage to the relay when the load is switched off.

TIMERS

Timers come in many styles and with different operating characteristics. Mechanical timers use either clock motors to operate a mechanical trip mechanism or solenoids to create timing operations. Pneumatic timers use air, a diaphragm, and an operating solenoid to cause a time delay.

FIGURE 10-9 Synchronous clock timer.

Figure 10–9 shows a mechanical time-clock type of timer. This type of timer is used for rough time-of-day timing to turn lights or equipment on or off at an approximate time of day. Tabs on the face of the clock are moved to create different on and off periods.

Pneumatic timers such as that in Figure 10–10 were typically used for "on" or "off" delay operations. The symbols shown in Figure 10–11 depict the type of timer operation and the operation of the contacts. In on-delay timers (time delay on energization, TDE), the contacts stay in their original position until a solenoid plunger has moved through its entire travel distance. The travel time is controlled by adjusting a needle valve to allow the air to escape in front of the solenoid diaphragm. The timer solenoid has power applied to pull the plunger into its timed-out position. The contacts can start as either normally open or normally closed and will change to the opposite position at the end of the time, if power is still applied to the solenoid coil.

FIGURE 10–10 Mechanical timer uses air (pneumatic) timing.

	TIMING SYMBOLS	
NOTC:	**On-Delay Timed Closed Contact**—*Timer contact normally open—timed closed upon relay energization—opens immediately upon relay de-energization.*	
NCTO:	**On-Delay Timed Open Contact**—*Timer contact normally closed—timed open upon relay energization—closes immediately upon relay de-energization.*	
NO TO:	**Off-Delay Timed Open Contact**—*Timer contact normally open— immediately closes upon relay energization—timer contact times open upon relay de-energization.*	
NCTC:	**Off-Delay Timed Closed Contact**—*Timer contact normally closed—opens immediately upon relay energization and timer contact times closed upon relay de-energization.*	
NO:	**Instantaneous Contact**—*Normally open—Upon relay energization, this contact closes immediately and opens immediately upon relay de-energization.*	
NC:	**Instantaneous Contact**—*Normally closed—Upon relay energization, this contact opens immediately and closes immediately upon relay de-energization.*	

FIGURE 10–11 Schematic symbols and explanations for various timing possibilities.

The off-delay timer (time delay on de-energization, TDD) acts in the opposite mode. In other words, the timing change to the contacts takes place when the power is removed from the timer coil and the solenoid plunger is allowed to go back to its de-energized position. The time that it takes the solenoid to return to its original position is controlled by the needle valve allowing air to return to the vacuum side of the air diaphragm. Figure 10–11 shows the TDD contacts in their normal, timed-out state. When power is applied to the timing coil in this type of timer, the contacts change to the opposite state and time out to return to the original designation.

Solid-state timers are now commonly used to provide highly accurate and an extremely wide range of timing operations. Many of the timers, such as the one in Figure 10–12, can be used as on-delays or off-delays and have several trigger modes. They can also provide a wide range of timing, from milliseconds to hundreds of hours. The timing is accomplished with electronics, which makes them accurate and precise.

Electronic timers come in four basic design capacities. *On-delay* timers provide timing functions just as the electromechanical timers do, but with a higher degree of accuracy

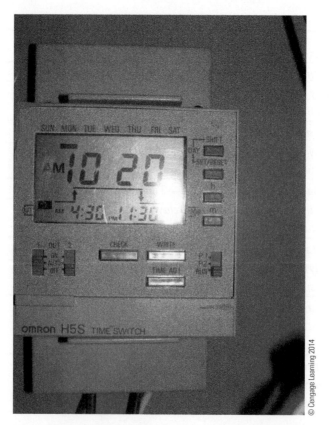

© Cengage Learning 2014

FIGURE 10–12 Individual electronic timers with a wide range of functions and timing features.

and repeatability. *Off-delay* timers also provide the timing function after the control signal is removed, as does the electromechanical timer. *Interval timers* are also called *one-shot timers*. With this timing sequence, the contact closure, whether dry contact or electronic conduction, occurs immediately after the control signal is applied. After a preset time, the contact reverts to its original condition and the timer waits for another signal to change its output again. The last type of timer is called a *recycler timer*. This timer cycles its output contact open and closed for preset intervals until the control signal is removed from the input. Symmetrical timers have equal on and off times for the output contacts as long as the input signal is present. The asymmetrical timer has adjustable times for the on and off portion of the output closure. See Figure 10–13 for the various individual solid-state timers. Some electronic timers are able to provide all these various functions within a single relay. The relay is microprocessor driven and fully programmable.

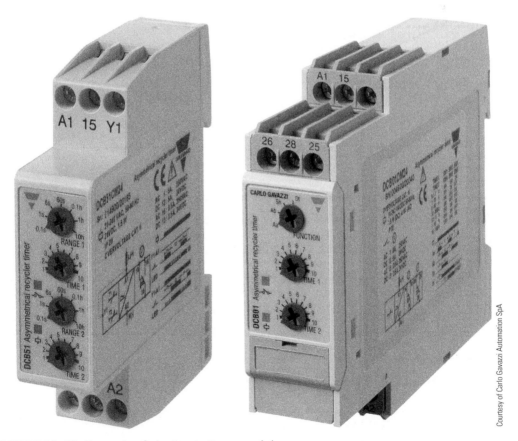

FIGURE 10–13 Example of electronic timer modules.

Courtesy of Carlo Gavazzi Automation SpA

SUMMARY

Relays are used to control various loads from other electrical circuits. They may be used to operate large values of DC or AC power or simply multiply the electrical circuit function. Contacts can be normally open (NO) or normally closed (NC) or convertible from one form to another.

Things to consider when ordering include

- control voltage and value.
- contact ratings in current and voltage, number of contacts, NO, NC, or SPDT, and so on.
- whether open contacts, enclosed contacts, or solid-state, no-contact movement are important.
- whether the relay needs very fast speed as in SSRs or whether physical isolation is more important as in electromechanical relays.

ACHIEVEMENT REVIEW

1. A mousetrap may be compared with the gate trigger action of an SCR. Indicate whether this statement is true or false, and explain your reasoning. _____

_____

_____

_____

_____

2. What is the major difference between an electromechanical relay and an SSR? _____

_____

_____

_____

_____

_____

3. Describe various relay control and load conditions.

_____

_____

_____

_____

4. How are relays used in industrial controls? _____

_____

_____

_____

_____

5. Basically, what is an SCR? _____

_____

_____

_____

_____

_____

UNIT 11

SUMMARY REVIEW
OF UNITS 5–10

OBJECTIVE

- To provide the student with an opportunity to evaluate the knowledge and understanding acquired in the study of the previous seven units.

A. Complete the following statements:

1. A DC motor starter is designed to limit the _____.

2. A three-terminal starting rheostat provides _____ protection for a shunt motor.

3. Because the holding coil of a four-terminal rheostat is connected across the source, this type of rheostat provides _____ protection to a motor.

4. A drum controller is used when the operator has _____ control of the motor.

B. Select the correct answer for items 5 through 13, and place the corresponding letter in the space provided.

5. Interpretation of automatic circuits requires the recognition of _____
   a. color.
   c. parallel circuits.
   b. rating.
   d. electrical circuit symbols.

6. A piece of apparatus that contains a minimum of one set of contacts operated by a coil is called a _____
   a. motor.
   c. relay.
   b. magnet.
   d. dynamo.

7. Normally closed contacts are _____
   a. open at all times.
   b. open when the relay coil is de-energized.
   c. open when the relay coil is energized.
   d. closed when the relay coil is energized.

8. A schematic automatic motor controller circuit diagram shows the _____
   a. actual wiring layout.
   b. schematic motor diagram.
   c. actual sequence of operations of the entire circuit.
   d. sequence of operations of the starting circuit only.

9. One type of automatic controller operates on the basis that the CEMF generated in a motor _____
   a. increases the starting current.
   b. energizes a starting relay.
   c. de-energizes a starting relay.
   d. reduces the field current.

10. Another type of controller accelerates the motor in steps by _____
    a. using a series of thermal elements.
    b. allowing starting resistor voltage drops to energize relays.
    c. operation of relays controlled by speed.
    d. cam-controlled relays.

11. A remote-controlled motor reversal requires _____
    a. one relay with two contacts.
    b. two single contactor relays.
    c. two double contactor relays.
    d. two interlocked relays.

12. The principle of dynamic braking involves _____
    a. the use of a dynamo as a brake.
    b. the use of an electromagnetic brake connected across the motor armature.
    c. connecting a low resistance across the motor armature.
    d. connecting a brake relay across the armature.

13. SCADA refers to a system that _____
    a. scans and records information.
    b. supervises control and data acquisition.
    c. is a protocol for machine interaction.
    d. operates to provide safety for machinery.

14. When humans need to interface with an automated machinery control, a typical method would be to use _____
    a. a microcontroller.
    b. multiple-screen PCs.
    c. an HMI.
    d. an MHI.

C. Complete the statements in items 15 through 19 by selecting the letter of the appropriate phrase or phrases from the list below right. Write the letter(s) in the space provided.

15. Undervoltage protection is provided in _____.

16. Dynamic braking is used for _____.

17. Multiple-step acceleration is used for starting _____.

18. Automatic starters provide more uniform acceleration
than _____.

19. A sealing circuit is used in _____.

a. quick stopping
b. large motors
c. a three-terminal
   starting rheostat
d. a four-terminal
   starting rheostat
e. small motors
f. series motors
g. manual starters
h. control circuits

20. Match each of the following symbols with its description at the right. Place the letter of
the symbol in the space provided.

a. ⊣⊢

b. ⊥

c. —Ⓜ—

d. ⚬⊥⚬

e. ⊬⊢

f. ⚬⋈⚬

g. ⊣▢⊢

© Cengage Learning 2014

1. _____ Normally closed pushbutton

2. _____ Thermal motor overload relay

3. _____ Shunt relay coil

4. _____ Normally open contact

5. _____ Fuse

6. _____ Normally closed contact

7. _____ Normally open pushbutton

UNIT  12

THE THREE-PHASE AC INDUCTION MOTOR

After studying this unit, the student should be able to

- describe the construction of a three-phase AC motor, listing the main components of this type of motor.

- identify the following items, and explain their importance to the operation of a three-phase AC induction motor: rotating stator field, synchronous speed, rotor-induced voltages, speed regulation, percent slip, torque, starting current, no-load power factor, full-load power factor, reverse rotation, and speed control.

- calculate motor speed and percent slip.

- reverse a squirrel-cage motor.

- describe why a motor draws more current when loaded.

- draw diagrams showing the dual-voltage connections for 240/480-volt motor operation.

- explain motor nameplate information.

OPERATING CHARACTERISTICS

The three-phase AC induction motor is relatively small in physical size for a given horsepower rating when compared with other types of motors. The squirrel-cage induction motor has very good speed regulation under varying load conditions. Because of its rugged construction and reliable operation, the three-phase AC induction motor is widely used for many industrial applications (Figure 12–1).

© Cengage Learning 2014

FIGURE 12–1 Three-phase motors used for a pumping application.

CONSTRUCTION DETAILS

The three-phase AC induction motor normally consists of a stator, a rotor, and two end shields housing the bearings that support the rotor shaft.

A minimum of maintenance is required with this type of motor because

- The rotor windings are shorted to form a squirrel cage.
- There are no commutator or slip rings to service (compared to the DC motor).
- There are no brushes to replace.

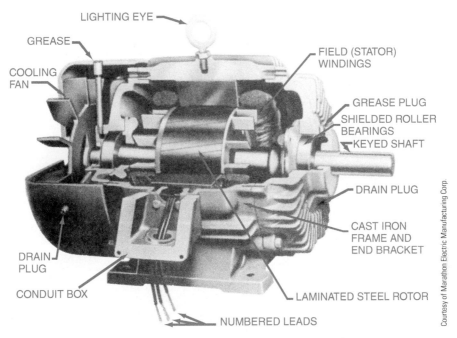

LIGHTING EYE

GREASE

COOLING FAN

FIELD (STATOR) WINDINGS

GREASE PLUG

SHIELDED ROLLER BEARINGS

KEYED SHAFT

DRAIN PLUG

CAST IRON FRAME AND END BRACKET

DRAIN PLUG

CONDUIT BOX

LAMINATED STEEL ROTOR

NUMBERED LEADS

Courtesy of Marathon Electric Manufacturing Corp.

FIGURE 12-2 Cutaway view of construction and features of a typical three-phase explosion-proof motor.

The motor frame is usually made of cast steel. The stator core is pressed directly into the frame. The two end shields housing the bearings are bolted to the cast steel frame. The bearings that support the rotor shaft are either sleeve bearings or ball bearings. Figure 12–2 is a cutaway view of an assembled motor. Figure 12–3 illustrates the main parts of a three-phase AC induction motor.

Stator

A typical stator contains a three-phase winding mounted in the slots of a laminated steel core (Figure 12–4). The winding itself consists of formed coils of wire connected so that three single-phase windings are spaced 120 electrical degrees apart. The three separate single-phase windings are then connected, usually internally, in either wye or delta. Three or nine leads from the three-phase stator windings are brought out to a terminal box mounted on the frame of the motor for single- or dual-voltage connections.

Rotor

The revolving part of the motor consists of steel punchings or laminations arranged in a cylindrical core (Figures 12–5, 12–6, and 12–7). Copper or aluminum bars are mounted near the surface of the rotor. The bars are brazed or welded to two copper end rings to form a short

FIGURE 12–3 Squirrel-cage rotor with internal fan and end bells of induction motor.

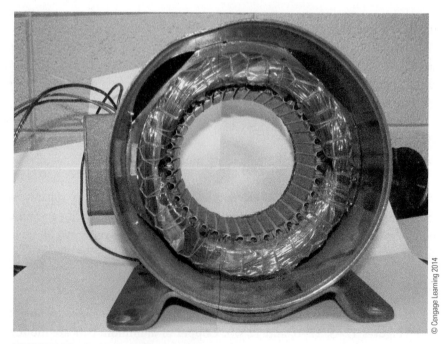

FIGURE 12–4 Stator of three-phase motor.

KEYWAY

Courtesy of General Electric Company

FIGURE 12–5 Squirrel-cage rotor for an induction motor.

Courtesy of General Electric Company

FIGURE 12–6 Cutaway view of a squirrel-cage rotor.

Courtesy of General Electric Company

FIGURE 12–7 Squirrel-cage form for an induction motor.

circuit for the rotor bars. In some small squirrel-cage induction motors, the bars and end rings are cast in one piece from aluminum.

Figure 12–5 shows such a rotor. Note that fins are cast into the rotor to circulate air and cool the motor while it is running. Note also that the rotor bars between the rings are skewed at an angle to the faces of the rings. Because of this design, the running motor is quieter and smoother in operation. A keyway is visible on the left end of the shaft. A pulley or load shaft coupling can be secured using this keyway.

Shaft Bearings

Typical sleeve bearings are shown in Figure 12–8 and Figure 12–9. The inside walls of the sleeve bearings are made of a babbitt metal that provides a smooth, polished, and long-wearing surface for the rotor shaft. A large, oversized oil slinger ring fits loosely around the rotor shaft and extends down into the oil reservoir. This ring picks up and slings oil on the rotating shaft and bearing surfaces. Two oil rings are shown in Figure 12–10. This lubricating oil

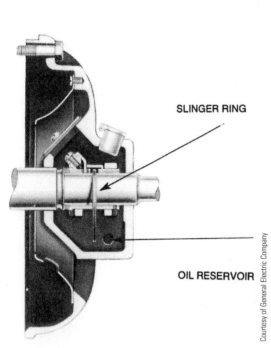

SLINGER RING

OIL RESERVOIR

Courtesy of General Electric Company

DRAIN PLUG

Courtesy of General Electric Company

FIGURE 12–8 Sleeve-bearing end shield for an open polyphase motor.

FIGURE 12–9 Sleeve-bearing end shield for a polyphase induction motor.

FIGURE 12–10 Partially assembled sleeve bearing for a totally enclosed 1250 hp motor.

film minimizes friction losses. An oil inspection cup on the side of each end shield enables maintenance personnel to check the level of the oil in the sleeve bearing.

Figures 12–11, 12–12, 12–13, and 12–14 illustrate ball bearing units. In many motors, ball bearings are used instead of sleeve bearings. Grease rather than oil is used to lubricate ball bearings. This type of bearing usually is two-thirds full of grease at the time the motor is assembled. Special fittings are provided on the end bells so that a grease gun can be used to apply additional lubricant to the ball bearing units at periodic intervals.

When lubricating roller bearings, remove the bottom plug so that the old grease is forced out. Consult the manufacturer's specifications for the motor for the lubricant grade recommended, the lubrication procedure, and the bearing loads.

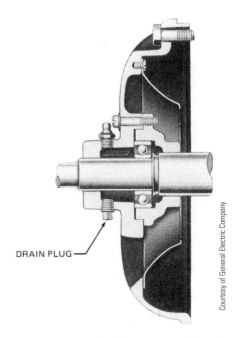

DRAIN PLUG

FIGURE 12–11 Ball-bearing end shield for an open polyphase motor.

FIGURE 12–12 Cutaway section of a single-row ball bearing.

FIGURE 12–13 Simple sealed-type of ball bearing.

FIGURE 12–14 Double-row ball bearing.

PRINCIPLE OF OPERATION OF A SQUIRREL-CAGE MOTOR

As stated in the previous discussion of stator construction, the slots of the stator core contain three separate single-phase windings. When three currents 120 electrical degrees apart pass through these windings, a rotating magnetic field results. This field travels around the inside of the stator core. The speed of the rotating magnetic field depends on the number of stator poles and the frequency of the power source. This speed is called the *synchronous speed* and is determined by the following formula:

$$\text{RPM} = \frac{120 \times f}{p} = \frac{120 \times 60}{6} = 1200 \text{ RPM}$$

RPM = Synchronous speed

f = Hertz (frequency)

p = Number of stator poles per phase

● **Example 1:** If a three-phase AC induction motor has six poles on the stator winding and is connected to a three-phase, 60-hertz source, then the synchronous speed of the revolving field is 1200 revolutions per minute (RPM).

As this magnetic field rotates at synchronous speed, it cuts the copper bars of the rotor and induces voltages in the bars of the squirrel-cage winding. These induced voltages set up currents in the rotor bars, which in turn create a field in the rotor core. This rotor field reacts with the stator field to cause a twisting effect, or torque, which turns the rotor. The rotor always turns at a speed slightly less than the synchronous speed of the stator magnetic field. This means that the stator magnetic field will always cut the rotor bars. If the rotor were to turn at the same speed as the stator field, the stator field would not cut the rotor bars and there would be no induced voltage or torque.

Speed Regulation and Percent Slip

The squirrel-cage induction motor has very good speed regulation characteristics (the ratio of difference in speed from no load to full load). Speed performance is measured in terms of percent slip. The synchronous speed of the rotating field of the stator is used as a reference point. Recall that the synchronous speed depends on the number of stator poles and the operating frequency. Because these two quantities remain constant, the synchronous speed also remains constant. If the speed of the rotor at full load is deducted from the synchronous speed of the stator field, the difference is the number of revolutions per minute that the rotor slips behind the rotating field of the stator. The value is expressed as percent slip according to the following formula:

$$\text{Percent slip} = \frac{\text{Synchronous speed} - \text{Rotor speed}}{\text{Synchronous speed}} \times 100$$

● **Example 2:** If the three-phase AC induction motor used in Example 1 has a synchronous speed of 1200 RPM and a full-load speed of 1140 RPM, find the percent of slip.

Synchronous speed (Example 1) = 1200 RPM

Full-load rotor speed = 1140 RPM

$$\text{Percent slip} = \frac{\text{Synchronous speed} - \text{Rotor speed}}{\text{Synchronous speed}} \times 100$$

$$\text{Percent slip} = \frac{1200 - 1140}{1200} \times 100$$

$$\text{Percent slip} = \frac{60}{1200} \times 100 = 0.05 \times 100$$

$$\text{Percent slip} = 5\%$$

For a squirrel-cage induction motor, as the value of percent slip decreases toward 0%, the speed performance of the motor is improved. The average range of percent slip for squirrel-cage induction motors is 2% to 6%.

Figure 12–15 shows a speed curve and a percent slip for a squirrel-cage induction motor operating between no load and full load. The rotor speed at no load slips behind the synchronous speed of the rotating stator field just enough to create the torque required to overcome friction and windage losses at no load. As a mechanical load is applied to the motor shaft, the rotor tends to slow down. This means that the stator field (turning at a fixed speed) cuts the rotor bars a greater number of times in a given period. The induced voltages in the rotor bars increase, resulting in more current in the rotor bars and a stronger rotor field. A greater magnetic reaction between the stator and rotor fields exists, which causes a stronger twisting effect, or torque. This also increases stator current taken from the line. The motor is able to handle the increased mechanical load with very little decrease in the speed of the rotor.

Typical slip-torque curves for a squirrel-cage induction motor are shown in Figure 12–16. The torque output of the motor in pound-feet (lb-ft) increases linearly with an increase in the

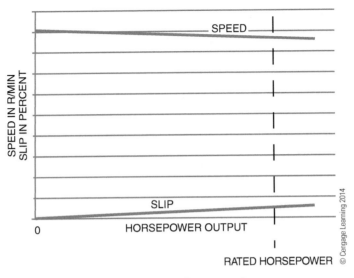

FIGURE 12–15 Speed curve and percent slip curve.

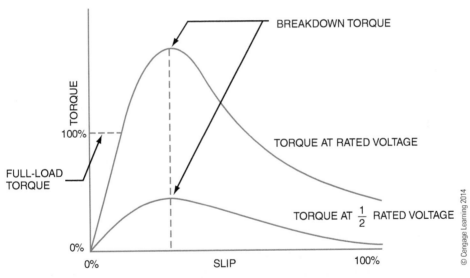

FIGURE 12–16 Slip-torque curves for a running squirrel-cage motor.

value of percent slip as the mechanical load is increased to the point of full load. Beyond full load, the torque curve bends and finally reaches a maximum point called the *breakdown torque*. If the motor is loaded beyond this point, there will be a corresponding decrease in torque until the point is reached where the motor stalls. However, all induction motors must have some slip in order to function. Starting torque is not shown, but is approximately 300% of running torque for a generic AC motor.

Starting Current

When a three-phase AC induction motor is connected across the full line voltage, the starting surge of current momentarily reaches as high as 400% to 1000% or more of the rated full-load current. At the moment the motor starts, the rotor is at a standstill. At this instant, therefore, the stator field cuts the rotor bars at a faster rate than when the rotor is turning. This means that there will be relatively high induced voltages in the rotor, which cause heavy rotor current. The resulting input current to the stator windings is high at the instant of starting. Because of this high starting current, starting protection rated as high as 300% of the rated full-load current for nontime-delay fuses is provided for squirrel-cage induction motor installations.

Most squirrel-cage induction motors are started at full voltage. If there are any questions concerning starting large motors at full voltage, the electric utility company should be consulted. In the event that the feeders and protective devices of the electric utility are unable to handle the large starting currents, reduced voltage starting circuits must be used with the motor. (See Unit 14.)

Power Factor

The power factor of a squirrel-cage induction motor is poor at no-load and light-load conditions. At no load, the power factor can be as low as 15% lagging. However, as load is applied to the motor, the power factor increases. At the rated load, the power factor may be as high as 85% to 90% lagging.

The power factor at no load is low because the magnetizing component of input current is a large part of the total input current of the motor. When the load on the motor is increased, the in-phase current supplied to the motor increases, but the magnetizing component of current remains practically the same. This means that the resultant line current is more nearly in phase with the voltage and the power factor is improved when the motor is loaded, compared with an unloaded motor.

Figure 12–17 shows the increase in power factor from a no-load condition to full load. In the no-load diagram, the in-phase current (I_W) is small when compared to the magnetizing current I_M; thus, the power factor is poor at no load. In the full-load diagram, the in-phase current has increased while the magnetizing current remains the same. As a result, the angle of lag of the line current decreases and the power factor increases.

Reversing Rotation

The direction of rotation of a three-phase induction motor can be readily reversed. The motor rotates in the opposite direction if any two of the three line leads are reversed (Figure 12–18). The leads are reversed at the motor.

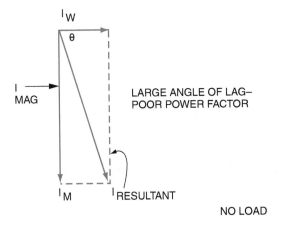

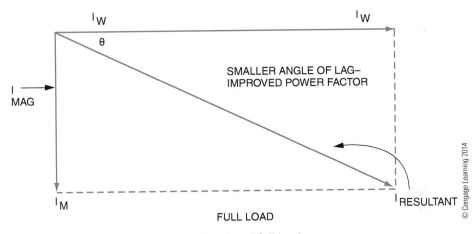

LARGE ANGLE OF LAG–
POOR POWER FACTOR

NO LOAD

SMALLER ANGLE OF LAG–
IMPROVED POWER FACTOR

FULL LOAD

FIGURE 12-17 Power factor at no load and full load.

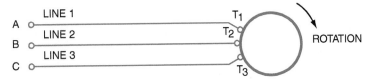

ROTATION BEFORE CONNECTIONS ARE CHANGED

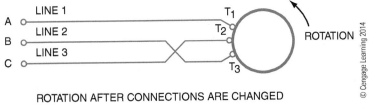

ROTATION AFTER CONNECTIONS ARE CHANGED

FIGURE 12-18 Reversing rotation of an induction motor.

© Cengage Learning 2014

Speed Control

A squirrel-cage induction motor has almost no speed variations without external controls. Recall that the speed of the motor depends on the frequency of the three-phase source and the number of poles of the stator winding.

The frequency of the supply line is usually 60 hertz in the United States, and is maintained at this value by the local power utility company. Because the number of poles in the motor is also a fixed value, the synchronous speed of the motor remains constant. As a result, obtaining a range of speed without changing the applied frequency is impossible. It can be controlled by a variable frequency AC electronic drive system or by changing the number of poles using external controllers. See Unit 14 on electronic frequency control.

INDUCTION MOTORS WITH DUAL-VOLTAGE CONNECTIONS

Many three-phase AC induction motors are designed to operate at two different voltage ratings. For example, a typical dual-voltage rating for a three-phase motor is 240/480 volts.

Figure 12–19 shows a typical wye-connected stator winding that can be used for either 240 volts, three phase, or 480 volts, three phase. Each of the three single-phase windings consists of two coil windings. Nine leads are brought out externally from this type of stator winding. These leads, identified as leads 1 to 9, end in the terminal box of the motor. To mark the terminals, start at the upper-left terminal T_1 and proceed in a clockwise direction in a spiral toward the center, marking each lead as indicated in the figure.

Figure 12–20 shows the connections required to operate a motor from a 480-volt, three-phase source. The two coils of each single-phase winding are connected in series. Figure 12–21 shows the connections to permit operation from a 240-volt, three-phase source.

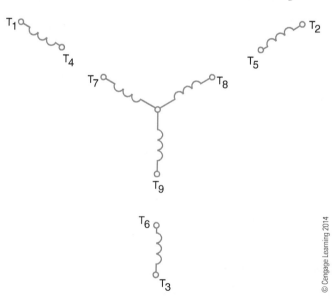

© Cengage Learning 2014

FIGURE 12–19 Method of identifying terminal markings.

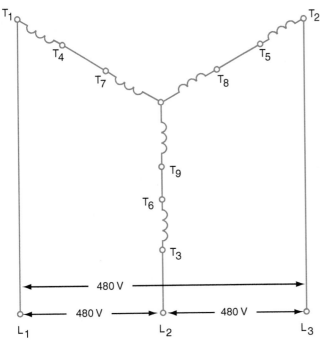

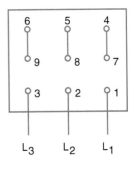

FIGURE 12-20 480-volt wye connection; coils are connected in series.

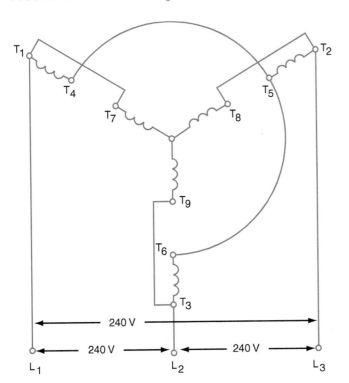

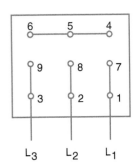

FIGURE 12-21 240-volt wye connection; coils are connected in parallel.

© Cengage Learning 2014

Wye- or Star-Connected Motors

If the lead identifications of a nine-lead (dual-voltage), three-phase, wye-connected motor have been destroyed, the electrician must reidentify them before connecting the motor to the line. The following method may be used. First, identify the internally connected common point by checking for continuity between three of the leads as in Figure 12–22(A).

Then identify the three other sets of coils by continuity between two leads at a time, shown in Figure 12–22(B). Assign T_7, T_8, and T_9 to any of the three leads of the permanent star-connected coils (A). Apply the lower-rated line voltage for the motor to T_7, T_8, and T_9, and operate to check the direction of rotation. Disconnect line voltage and connect one of the undetermined coils to T_7. Reconnect power, leaving the lines on T_7, T_8, and T_9. If the coil is correctly connected and is the proper coil, the voltage should be about 1.5 times the line voltage between the loose end and the other two lines. *Be careful of line voltage.*

If the correct coil is selected but reversed, the voltage between the loose end and the other two leads will be about 58% of the line voltage. If the wrong coil is selected, the voltage differences between the loose end and the other two line leads will be uneven. See Figure 12–22(C).

When the readings are even and approximately 1.5 times the line voltage, mark the lead connected to T_7 as T_4 and other end of the coil as T_1.

Perform the same tests with another coil connected to T_8. Mark these leads T_5 and T_2. Perform the same test with the last coil connected to T_9 to identify the T_3 and T_6 leads.

Connect L_1 to T_1, L_2 to T_2, L_3 to T_3, T_4 to T_7, T_5 to T_8, T_6 to T_9, and then operate the motor. The motor should operate quietly in the same direction as before.

Delta-Connected Motors

Another connection pattern for three-phase motors is the delta-connected motor. It is so named because the resulting schematic pattern looks like the Greek letter delta (Δ). A method of identifying and connecting these leads is necessary because this pattern is different from the star- or wye-connected motor.

Properly connecting the leads of a delta-connected, three-phase, dual-voltage motor presents a problem if the lead markings are destroyed.

First, the electrician must determine whether the motor is delta connected or star connected. Both motors have nine leads if they are dual-voltage motors. However, the delta- connected motor has three sets of three leads that have continuity, and the star-connected motor has only one set of three.

After it is determined that the motor is delta connected, a sensitive ohmmeter is needed to find the middle of each group of three leads. The ohm values are low when using the DC power of an ohmmeter, so use care in identifying the center of each coil group. Label the center of each

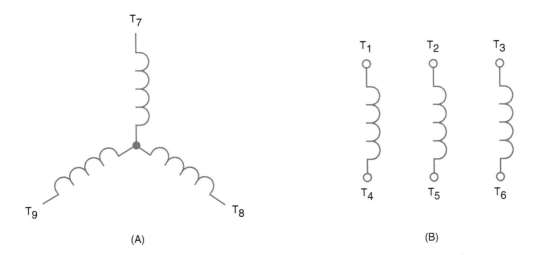

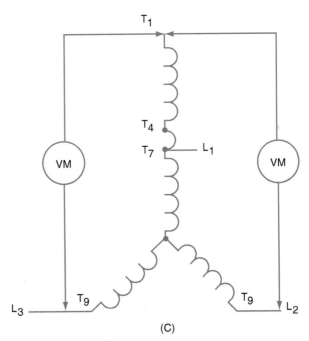

FIGURE 12-22 Wye- or star-connected motor. (A) Internal star-point connection. (B) Coil group lead marking. (C) Checking for proper coil lead markings on wye-connected, dual-voltage motor.

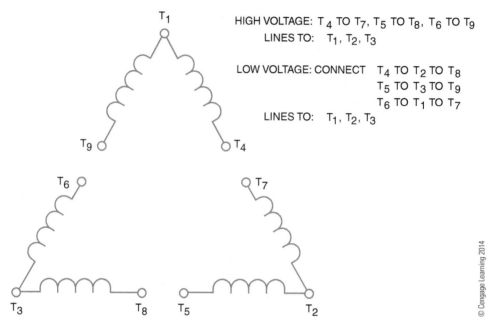

FIGURE 12–23 Nine leads of a delta-connected, three-phase, dual-voltage motor.

group T_1, T_2, and T_3, respectively. Using masking tape, temporarily label the other leads of the T_1 group as T_4 and T_9 (see Figure 12–23).

Temporarily mark the ends of the T_2 group as T_5 and T_7, and mark the ends of the T_3 group as T_6 and T_8.

Connect the lower motor voltage rating using lines 1, 2, and 3 to T_1, T_4, and T_9. The other coils will have induced voltage, so be careful not to touch the other loose leads to each other or to yourself!

Disconnect the power and connect the lead marked T_4 to T_7. Reconnect the power as before and read the voltage between T_1 and T_2. If the markings are correct, the voltage should be about twice the applied line voltage. If it reads about 1.5 times the line voltage, reconnect T_4 to the lead marked T_5. If the voltage T_1 to T_2 then goes to line voltage, reconnect T_9 to T_7, thereby reversing both coils. When the voltage T_1 to T_2 equals twice the applied line voltage, mark the leads connected together as T_4 from the T_1 group connected to T_7 of the T_2 group.

Now use the third coil group. Leave the lower line voltage connected to the first group as before. Test and connect the leads so that when T_9 is connected to a lead of the third group, the T_1-to-T_3 voltage is twice the applied line voltage. Mark the lead connected to T_9 as T_6 and the other end of the coil group as T_8.

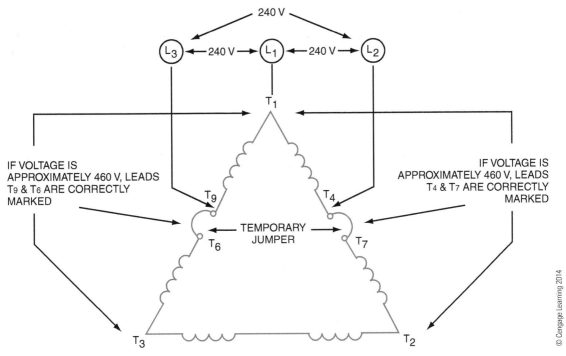

FIGURE 12-24 Illustration of voltage tests used to determine correct lead markings on a delta motor.

To double-check, disconnect the line lead from T_9 and reconnect it to T_7; disconnect the line lead from T_1 and reconnect it to T_2; disconnect the line lead from T_4 and reconnect it to T_5. The motor should run in the same direction as before. If it does not, recheck the lead markings.

To check further, move the line leads from T_7 to T_8, from T_5 to T_6, and from T_2 to T_3. Start the motor. Rotation should be the same as in the previous steps. *Be careful! Voltage is induced into other windings.* (See Figure 12–24.)

MOTOR NAMEPLATES

Motor nameplates provide information vital to the proper selection and installation of the motor. Most useful data given on the nameplate refer to the electrical characteristics of the motor. Given this information and using the *National Electrical Code*®, the electrician can determine the conduit, wire, and starting and running protection sizes. (The *NEC*® provides minimum requirements.)

The design and performance data on the nameplate are useful to maintenance personnel. The information is vital for the fast and proper replacement of the motor, if necessary.

MANUFACTURER'S NAME INDUCTION MOTOR MADE IN U.S.A.		
SERIAL NO.	TYPE	MODEL
HP	FRAME	SV. FACTOR
AMPS	VOLTS	INSUL.
RPM	HERTZ	kVA
DUTY	PHASE	TEMP C°
NEMA NOM. EFF.	dBA/NOISE	THERMAL-PROTECTED SEALED BEARINGS

© Cengage Learning 2014

FIGURE 12-25 Typical motor nameplate.

For a better understanding of the motor, typical information found on motor nameplates is described here (Figure 12–25):

- The manufacturer's name can be found on the nameplate.

- *Type* identifies the type of the enclosure. This is the manufacturer's coded identification system.

- *Serial number* is the specific motor identification. This is the individual number assigned to the motor, similar to a social security number for a person. It is kept on file by the manufacturer.

- The *model number* is an additional manufacturer's identification, commonly used for ordering purposes.

- *Frame size* identifies the measurements of the motor.

- *Service factor* (or SF). An SF of 1.0 means the motor should not be expected to deliver more than its rated horsepower. The motor will operate safely if it is run at the rated horsepower times the SF, maximum. Common SFs are 1.0 to 1.15. It is recommended that the motor not be run continuously in the SF range. This may shorten the life expectancy of the insulation system.

- *Amperes* means the current drawn from the line when the motor is operating at rated voltage and frequency at the fully rated nameplate horsepower.

- *Volts* should be the value measured at the motor terminals and for which the motor is designed.

- The *class of insulation* refers to the insulating material used in winding the motor stator. For example, in a Class B system, the maximum operating temperature is 130°C

(266°F); for Class F, it is 155°C (311°F); and for Class H, it is 180°C (356°F). This is the maximum temperature of the motor coil.

- *RPM* means the speed in revolutions per minute when all other nameplate conditions are met.

- *Hertz* is the frequency of the electric power system for which the motor is designed. Performance will be altered if it is operated at other frequencies.

- *Duty* is the cycle of operation at which the motor can safely operate. "Continuous" means that the motor can operate fully loaded 24 hours a day. If "intermediate" is shown, a time interval also appears. This means the motor can operate at full load for the specified period. The motor should then be stopped and allowed to cool before starting again.

- *Ambient temperature* specifies the maximum surrounding air temperature at which the motor can operate to deliver the rated horsepower.

- *Phase* indicates the number of voltage phases at which the motor is designed to operate.

- *kVA* is a code letter that indicates the locked rotor kVA per horsepower. This is used to determine starting equipment and protection for the motor. A code letter table is found in *NEC® 430.7(B)*.

- *Efficiency* is expressed as a percentage. This value is found on standard motors as well as "premium efficiency" motors.

- *Noise.* Some motors are designed for low noise emission. The noise level given on the nameplate is measured in "dBA" units of sound.

- *Manufacturer's notes* list specific features of the motors, such as "thermal protected" and/ or "scaled bearings."

- *Design letter* is a manufacturer's coded letter for the motor design that affects operating characteristics. Five standard classifications are designs A, B, C, D, and E. Each design has a special speed versus torque curve and is used for different applications. Design B is the most common.

ALTITUDE

Manufacturers' guarantees for standard motor ratings are usually based on operation at any altitude up to 3300 ft. Motors suitable for operation at an altitude higher than 3300 ft above sea level are of special design and/or have a different insulation class. For example, standard motors with an SF of 1.15 may be operated up to an altitude of 9900 ft by reducing the SF. At an altitude of 9900 ft, the SF would be 1.00. It may be necessary to derate the motor or use a larger frame size.

SUMMARY

Three-phase AC induction motors use a squirrel-cage winding in the rotor. There are no electrical connections to the rotor, but current is induced into the rotor windings by electromagnetic induction. The squirrel-cage winding produces a magnetic field that is pushed and pulled by the stator magnetic field.

     The rotor is supported by a steel shaft that must rotate. The shaft is allowed to rotate with the application of different types of bearings and various lubrications. Synchronous speed, speed regulation, and percent slip are all calculations used in determining the speed of the rotor. Motor electrical characteristics such as power factor and starting current are related to the electrical design of the motor.

     If motor lead markings are destroyed, the leads can be remarked according to the procedures outlined in this unit. Motor nameplate data are critical information to be used when ordering replacement motors. Some nameplate information is essential for proper replacement of the operating characteristics, and other data are used to size the electrical supply and the motor protection.

ACHIEVEMENT REVIEW

A.   Answer the statements and questions in items 1 through 8.

1.   List the essential parts of a squirrel-cage induction motor. _____

_____

_____

_____

_____

_____

2.   State two advantages of using a squirrel-cage induction motor. _____

_____

_____

3.   State two disadvantages of a squirrel-cage induction motor. _____

_____

_____

4. List the two factors that determine the synchronous speed of an induction motor.

_____

_____

_____

5. Explain how to reverse the direction of rotation of a three-phase AC induction motor.

_____

_____

_____

_____

6. A four-pole, 60-hertz, three-phase AC induction motor has a full-load speed of 1725 RPM. Determine the synchronous speed of this motor. _____

_____

_____

_____

_____

_____

7. What is the percent slip of the motor described in question 6?

_____

8. Why is the term *squirrel-cage* applied to this type of three-phase induction motor?

_____

_____

_____

B. Select the correct answer for the statements in items 9 through 14, and place the corresponding letter in the space provided.

9. Who or what determines whether large induction motors may be started at full voltage across the line? _____
   a. maximum motor size
   b. rated voltage
   c. the power company
   d. Department of Building and Safety

10. The power factor of a three-phase AC induction motor operating unloaded is _____
   a. the same as with full load.
   b. very poor.
   c. very good.
   d. average.

11. The power factor of a three-phase AC induction motor operating with a full load _____
   a. improves from no load.
   b. decreases from no load.
   c. remains the same as at no load.
   d. becomes 100%.

12. The squirrel-cage induction motor is popular because of its characteristics of _____
   a. high percent slip.
   b. low percent slip.
   c. simple, rugged construction.
   d. good speed regulation.

13. The speed of a squirrel-cage induction motor depends on _____
   a. voltage applied.
   b. frequency and number of poles.
   c. field strength.
   d. current strength.

14. Synchronous speed (RPM) is calculated using the formula _____
   a. $p = \dfrac{120 \times f}{\text{RPM}}$

   b. $\text{RPM} = \dfrac{120 \times f}{p}$

   c. $\text{RPM} = \dfrac{120 \times p}{f}$

   d. $\text{RPM} = \dfrac{120 \times f}{p}$

C. Draw the following connection diagrams.

15. Show the connection arrangement for the nine terminal leads of a wye-connected, three-phase motor rated at 240/480 volts for operation at 480 volts, three phase.

16. Show the connection arrangement for the nine terminal leads of a wye-connected, three-phase motor rated at 240/480 volts for operation at 240 volts, three-phase.

UNIT 13

STARTING THREE-PHASE AC INDUCTION MOTORS

OBJECTIVES

After studying this unit, the student should be able to

- state the purpose of an across-the-line magnetic starting switch.

- describe the basic construction and operation of an across-the-line starter.

- state the ratings for the maximum sizes of fuses required to provide starting protection for motors in the various code marking groups.

- describe what is meant by running overload protection.

- draw a diagram of the connections for an across-the-line magnetic starter with reversing capability.

AC motors do not require the elaborate starting equipment that must be used with DC motors. Most three-phase AC induction motors with ratings up to 10 horsepower are connected directly across the full line voltage. In many cases, motors with ratings greater than 10 horsepower also can be connected directly across the full line voltage. Across-the-line starting usually is accomplished using a magnetic starting switch controlled from a pushbutton station.

The electrician regularly is called upon to install and maintain magnetic motor starters. As a result, the electrician must be very familiar with the connections, operation, and troubleshooting of these starters. The *National Electrical Code®* provides information on starting and running overload protection for squirrel-cage induction motors. A comprehensive study of motor controls is provided in *Electric Motors and Motor Controls* (Delmar, Cengage Learning).

ACROSS-THE-LINE MAGNETIC STARTER

In the simplest starting arrangement, the three-phase AC motor is connected across full line voltage for operation in one direction of rotation. This is referred to as *across-the-line (ATL) starting*. The magnetic switch used for starting has three heavy contacts, one auxiliary contact, three motor overload relays, and an operating coil. The magnetic switch is called a *motor starter* if it has overload protection. Older motor starters already in service may have used two overload relays. Three overload relays are required by the *National Electrical Code®* in new installations.

The wiring diagram for a typical across-the-line magnetic starter is shown in Figure 13–1. The three heavy power contacts are in the three line leads feeding the motor. The auxiliary contact acts as a sealing circuit around the normally open start pushbutton when the motor is operating. As a result, the relay remains energized after the start button is released. The four contacts of the ATL magnetic starter are operated by the magnetic starter coil controlled from a pushbutton station, as shown in the schematic diagram in Figure 13–2.

Figure 13–3 shows a typical pushbutton station. Two pushbuttons are housed in a pressed steel box. The start pushbutton is normally open and the stop pushbutton is normally closed, as shown in the diagram (Figure 13–4).

STARTING PROTECTION (BRANCH-CIRCUIT, SHORT-CIRCUIT, AND GROUND-FAULT PROTECTION)

In Figure 13–1, a motor-rated disconnect switch is installed ahead of the magnetic starter. The safety switch is a three-pole, single-throw enclosed switch. It has a quick-break spring action and is operated externally. The motor circuit switch contains three cartridge fuses that serve as the short-circuit protection for the motor. These fuses must have sufficient capacity to handle the starting surge of current to the motor. The fuses protect the installation from possible

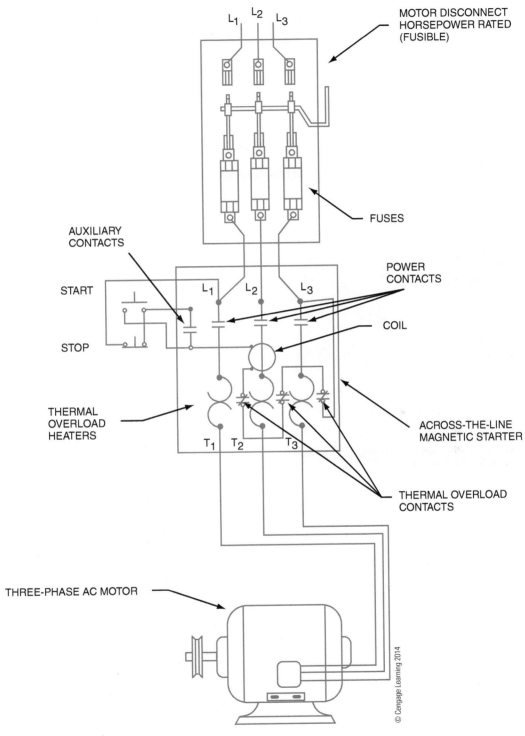

FIGURE 13-1 A wiring diagram for an ATL magnetic starter.

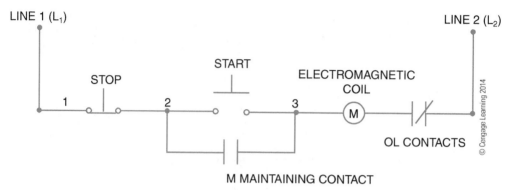

FIGURE 13–2 Elementary diagram of control circuit for the starter.

FIGURE 13–3 Internal wiring and cover of start/stop station.

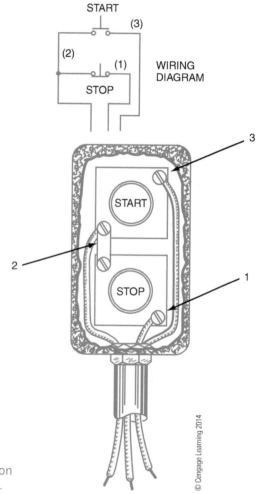

FIGURE 13–4 A pushbutton station and wiring diagram.

damage resulting from defective wiring or faults in the motor windings. This combination of switch and fuse protection and motor starter is available in a single enclosure (Figure 13–5). (See NEC® Article 430, Part IV.)

Briefly, the *National Electrical Code®* provides the following information on starting protection for squirrel-cage induction motors.

The maximum size of fuses permitted to protect motors are rated at 300% of the full-load current of the motor for nontime-delay fuses, and 175% for time-delay fuses.

Note: If the required fuse size as determined by applying the given percentages does not correspond with the standard sizes of fuses available, and if the specified overcurrent protection is not sufficient to handle the starting current of the motor, then the next higher standard fuse size may be used. In no case can the fuse size exceed 400% of the full-load current of the motor for nontime-delay fuses and 225% of the full-load current for time-delay fuses. (See NEC® Article 430, Part IV, Article 430.52(C)(1) Exceptions.)

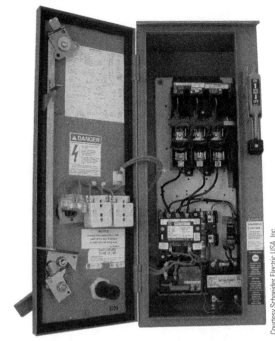

FIGURE 13–5 Combination starter with fusible disconnect switch.

Rotors are constructed with different characteristics. Figure 13–6 shows the various types of rotor construction and associated code letters. The applications of motors with these code letters are also indicated. The design of the rotor affects the amount of current needed to produce the rotor magnetic field. Code letter A has high starting torque and relatively low starting current. The NEC® Table 430.7(B) indicates that a code letter A motor will have less locked rotor kVA than other motors. This calculation indicates that there is less starting current for the same voltage for a code A compared to a code K motor. The chart in Figure 13–6 provides some broad categories of motors (A, B to E, and F to V).

An AC magnetic starter is shown in Figure 13–7. The starter consists of power contacts that are used to open and close the circuit to the motor. As AC is applied to the magnetic coil, the magnet draws the contacts closed and connects the line power to the motor power. In addition to connecting the line power, the magnetic starter has an add-on block at the bottom to provide for running overload protection. See Unit 14 for detailed operation of the magnetic starter.

© Cengage Learning 2014

INDUCTION MOTOR WITH CODE LETTER A

THIS TYPE OF MOTOR HAS A HIGH-RESISTANCE ROTOR WITH SMALL ROTOR BARS NEAR THE ROTOR SURFACE. THIS MOTOR HAS A HIGH STARTING TORQUE AND LOW STARTING CURRENT.

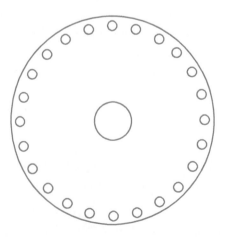

APPLICATIONS:

METAL SHEARS, PUNCH PRESSES, AND METAL DRAWING MACHINERY.

INDUCTION MOTOR WITH CODE LETTERS B TO E

THIS TYPE OF MOTOR HAS A HIGH-REACTANCE AND LOW-RESISTANCE ROTOR. THIS MOTOR HAS A RELATIVELY LOW STARTING CURRENT AND ONLY FAIR STARTING TORQUE. IT HAS LARGER CONDUCTORS DEEP IN THE ROTOR IRON.

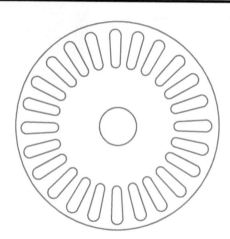

APPLICATIONS:

MOTOR-GENERATOR SETS, CENTRIFUGAL PUMPS, OR ANY APPLICATION WHERE A HIGH STARTING TORQUE IS NOT REQUIRED AND LOW STARTING CURRENT IS REQUIRED.

INDUCTION MOTOR WITH CODE LETTERS F TO V

THIS TYPE OF MOTOR HAS A RELATIVELY LOW-RESISTANCE AND LOW-INDUCTIVE REACTANCE ROTOR. THIS MOTOR HAS A HIGH STARTING CURRENT AND ONLY A FAIR STARTING TORQUE. IT HAS LARGE CONDUCTORS NEAR THE ROTOR SURFACE.

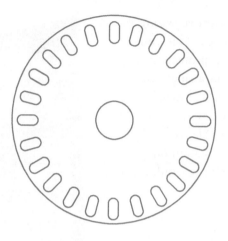

APPLICATIONS:

FANS, BLOWERS, CENTRIFUGAL PUMPS, OR ANY APPLICATION WHERE A HIGH STARTING TORQUE IS NOT REQUIRED AND HIGH STARTING CURRENT IS TOLERATED.

FIGURE 13–6 Various types of rotor laminations.

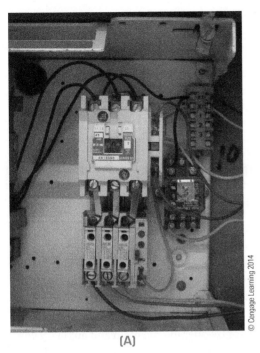

© Cengage Learning 2014

(A)

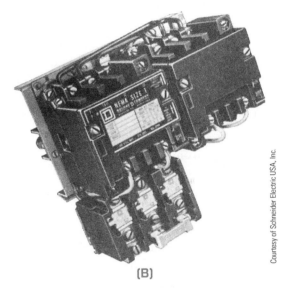

Courtesy of Schneider Electric USA, Inc.

(B)

FIGURE 13-7 (A) A magnetic contactor and overload section make up a magnetic starter. (B) AC reversing-magnetic motor starter. The elementary diagram of the starter is shown in Figure 13–10.

● *Example 1:* A three-phase AC induction motor with a nameplate marking of code letter F is rated at 5 hp, 230 volts. According to the *National Electrical Code®*, this motor has a full-load current per terminal of 15.2 amperes according to *Table 430.250*. The starting protection shall not exceed 300% of the rated current for squirrel-cage motors with nontime-delay fuses. Thus, the starting protection is 15.2 × 3 = 45.6 amperes.

Because a 45.6-ampere fuse cannot be obtained (see *NEC® 240.6*), the next larger size of fuse (50 amperes) should be used. For motor branch-circuit protection, the motor current listed in the appropriate table of the *National Electrical Code®* should be used. The full-load current, as stated on the motor nameplate, is not used for this purpose.

RUNNING OVERLOAD PROTECTION (NEC®: MOTOR AND BRANCH-CIRCUIT OVERLOAD PROTECTION)

Many motor starters installed in the United States use a thermal type of overload assembly. The assembly is normally located beneath the contactor and is directly attached to the magnetic contactor. The overload monitoring system is designed to measure the amount of current flowing to the motor through the contactor. This is done by connecting thermal sensors called *heaters* in series with the motor current. The heaters are sized to produce a certain amount of heat with a specified current through them. They are calibrated to cause a thermally operated switch to open when there is sustained heat. The extra heat is caused by too much current flow to the motor, which indicates the motor is jammed or is working too hard and is overloaded. The thermal sensors are various types as seen in Figure 13–8. The heater sensors with the associated trip-overload relays are pictured. The *National Electrical Code*® requires the use of three overload units as running-overload protection *Table 430.37*. Although new installations require three overload relays, electricians work on many older installations that have only two overload relays. These were installed before the three overload relay requirement became effective. The overload relay unit may be either three individual units or a common block containing the three heaters and only one trip switch contact unit reacting from any one of the heaters.

These overload sensing units are made of a special alloy. Motor current through these units causes heat to be generated. In one type, a small bimetallic strip is located next to each of the three heater units. When an overload on a motor continues for a period of approximately 1 to 2 minutes, the excessive heat developed by the heater units causes the bimetallic strips to expand. As each bimetallic strip expands, it causes the normally closed contacts in the control circuit to open. The main relay coil is de-energized and disconnects the motor by opening the main and auxiliary contacts. Melting alloy overloads (solder pots) also are commonly used. The heat generated by the overload melts the solder pot to release a ratchet that trips the control circuit contacts.

Before the motor can be restarted at the pushbutton station, the overload contacts in the control circuit *must be allowed to cool* before being reclosed (reset). When the reset button in the magnetic starter is pressed, the overload contacts in the control circuit are reset to their normally closed position. The motor then can be controlled from the pushbutton station.

The *National Electrical Code*® requires that the running overload protection in each phase be rated at not more than 125% of the nameplate full-load current rating for motors that are marked with a temperature rise of 40°C (104°F) or less (see *NEC*® *Article 430, Part III*).

© Cengage Learning 2014

FIGURE 13–8 Left is solder pot heater, middle is just heater for melting alloy OL, right is heater for bi-metal OL.

● *Example 2:*   Using the motor full-load current rating from the nameplate data, determine the running overcurrent protection for a three-phase, 5 hp, 230-volt
AC induction motor with a rated full-load current of 14.5 amperes and a
temperature rise of 40°C (104°F). The running overcurrent protection is
14.5 × 1.25 = 18.1 amperes.

For this motor, heater overload units rated to trip at 18.1 amperes are chosen for the magnetic starter. Where the overload relay so selected is not sufficient to start this motor, the next
higher size overload relay is permitted, but not to exceed 140% of the motor full-load current
rating [*NEC*® *Article 430.32(C)*]. Actual motor nameplate currents are used to establish the
overload protection.

HOW TO SIZE OVERLOAD THERMAL PROTECTION

The process of determining the correct overload protection to protect a motor from excessive
heating due to mechanical work overloads or from failure to start is a matter of following all
the rules of the *NEC*®. Also, each manufacturer has methods to determine proper protection.
Article 430, Part III, determines the parameters of protection. *Article 430.32* determines the trip
point and the percentage of full-load current when the overload must open the power circuit
to the motor. *Article 430.33* determines the overload protection for intermittent or similar non-
continuous duty motors. Once you have determined the actual trip percentage point according
to the *Code*, you multiply the actual motor nameplate current times the percentage to determine
the actual current trip value.

Most manufacturers use the system of labeling the overload heaters based on the nameplate information on the motor. Each manufacturer uses its own numbering system that corresponds to the motor starter used. If the motor is a standard rating with a service factor of 1.15 or
greater and the marked temperature rise for the motor installation is 40°C (104°F) or less, then
choose the heater catalog number that corresponds to the nameplate rating. This heater will
provide a trip point of approximately 125% of the nameplate rating as required by *NEC*® *Article
430.32(A)(1)*. However, if the service factor is less than 1.15, typically 1.0, or the motor has
a marker temperature rise over 40°C (104°F), typically 50°C (122°F), then the *Code* requires
closer tolerance, and you choose one size smaller from the manufacturer's table. The one size
smaller will yield 115% protection as required.

In the case of 125% or 115% protection, the *Code* does allow the electrical personnel
to increase the ratings if the original selection does not allow the motor to operate. *Article
430.32(C)* allows us to increase each category by one size provided the 125% category does not
exceed 140%, and the 115% category does not exceed 130%. All the above situations are based
on the fact that the controller and the motor are in the same ambient temperature.

When sizing the overload heater from a manufacturer's listing, you must determine the
heater catalog numbers. There are standard methods used to determine the heater number.
The electrical technician needs to determine whether the motor controller that contains the

overload heater is in the same ambient temperature as the motor it controls. If the controller is in a higher ambient temperature, that means the overload sensors are already warmer than the motor, and not as much heat can be added before they trip and disconnect the motor. Therefore, choose a heater with a higher number than originally selected. Conversely, if the control is at a lower ambient temperature than the motor, select a heater that is one size less than the original choice. The lower number of the heater produces more heat per ampere of current flow and compensates for the lower ambient temperature.

Electronic overload modules are sized according to the nameplate of the motor and then adjusted from 115% to 140% depending on the circumstances of the motor nameplate and the conditions relating to the location of the motor and controller. In the following description, the overload monitor can also monitor for single phasing of three-phase motors where one phase fails and the motor continues to run on the other two phases, now a single-phase supply. The three-phase motor is not designed to operate safely with only one phase and may be damaged if allowed to run too long. The electronic overload monitor causes the controller to disconnect the motor.

● **Example 3:** A 5 hp, 230 V, three-phase motor has a nameplate current of 14.5 A and a service factor of 1.25. The marked temperature rise is 40°C (104°F). The controller and the motor are at the same ambient temperature. Use the heater chart in Figure 13–9.

● **Solution:** If the controller and motor are at the same ambient temperature, the motor has a service factor of 1.15 or more, and a temperature rise of 40°C (104°F) or less, then simply choose 125% protection and use the nameplate rating of the motor to pick the heater catalog number H37 for NEMA size 00, 0,1 starters. You need three heaters for a conventional overload-heater–style starter for a three-phase motor.

● **Example 4:** A 5 hp, 230 V motor has a nameplate current of 14.5 A and a service factor of 1.0. The marked temperature rise is 40°C (104°F). The controller is in a cooler location than the motor by approximately 10°C (50°F). Use the heater chart in Figure 13–9.

● **Solution:** First choose the heater overload as if the controller and the motor were at the same temperature. In this case, because the service factor is 1.0, the *NEC*® requires us to protect the motor at 115% of the nameplate current. Use the chart to determine the heater selection at full nameplate current, and then choose one size smaller to satisfy the 115% requirement. Now adjust that size based on the ambient temperature of the controller. Because the controller is already cooler, choose one size smaller than normal. In this case, the heater is two sizes smaller than would be indicated by simply using the nameplate current. The heater is H36. If this does not allow the motor to run normally without overheating, the size can be adjusted up one size according to *Article 430.32(C)*.

Table 134

Size								Heater
00, 0, 1	1P, 1P¾	2, 2½	3, 3½	4 (JB)	4 (JG)	4½, 5	6	Code
0.35–0.39	0.35–0.39	—	—	—	—	—	—	H1
0.40–0.44	0.40–0.44	—	—	—	—	—	—	H2
0.45–0.50	0.45–0.50	—	—	—	—	—	—	H3
0.51–0.55	0.51–0.55	—	—	—	—	—	—	H4
0.56–0.61	0.56–0.61	—	—	—	—	—	—	H5
0.62–0.68	0.62–0.68	—	—	—	—	—	—	H6
0.69–0.74	0.69–0.74	—	—	—	—	—	—	H7
0.75–0.82	0.75–0.82	—	—	—	—	—	—	H8
0.83–0.92	0.83–0.92	—	—	—	—	—	—	H9
0.93–1.02	0.93–1.02	—	—	—	—	—	—	H10
1.03–1.13	1.03–1.13	—	—	—	—	—	—	H11
1.14–1.29	1.14–1.29	—	—	—	—	—	—	H12
1.30–1.45	1.30–1.45	—	—	—	—	—	—	H13
1.46–1.57	1.46–1.57	—	—	—	—	—	—	H14
1.58–1.74	1.58–1.74	—	—	—	—	91.5–101	183–203	H15
1.75–1.92	1.75–1.92	—	—	—	—	102–110	204–221	H16
1.93–2.18	1.93–2.18	—	—	—	—	111–125	222–251	H17
2.19–2.34	2.19–2.34	—	—	—	—	126–138	252–277	H18
—	—	—	—	—	—	139–146	278–293	H19
2.35–2.62	2.35–2.62	—	—	—	—	147–163	294–327	H20
2.63–2.97	2.63–2.97	—	—	—	—	164–175	328–351	H21
2.98–3.25	2.98–3.25	—	—	—	—	176–192	352–385	H22
3.26–3.60	3.26–3.60	—	—	—	—	193–213	386–427	H23
3.61–4.00	3.61–4.00	—	—	—	—	214–238	428–477	H24
4.01–4.50	4.01–4.50	—	—	—	—	239–248	478–525	H25
4.51–5.10	4.51–5.10	—	—	—	—	—	526–580	H26
—	—	—	—	—	—	—	—	H27
5.11–6.32	5.11–6.32	—	—	—	—	—	—	H28
6.33–6.81	6.33–6.81	—	—	—	—	—	—	H29
6.82–7.69	6.82–7.69	—	—	—	—	—	—	H30
7.70–8.54	7.70–8.54	—	—	—	—	—	—	H31
8.55–9.54	8.55–9.54	—	—	—	—	—	—	H32
9.55–10.7	9.55–10.7	—	—	19.1–20.4	—	—	—	H33
10.8–12.1	10.8–12.1	12.4–13.5	—	20.5–22.0	—	—	—	H34
12.2–12.7	12.2–12.7	13.6–14.3	—	22.1–23.6	—	—	—	H35
12.8–14.0	12.8–14.0	14.4–15.6	—	23.7–25.4	—	—	—	H36
14.1–14.9	14.1–14.9	15.7–17.8	—	25.5–27.3	—	—	—	H37
15.0–16.9	15.0–16.9	17.9–18.8	—	27.4–29.4	—	—	—	H38
17.0–18.1	17.0–18.1	18.9–19.9	—	29.5–31.6	—	—	—	H39
18.2–18.9	18.2–18.9	20.0–22.1	—	31.7–34.0	—	—	—	H40
19.0–22.0	19.0–22.0	22.2–25.5	—	34.1–36.6	—	—	—	H41
22.1–24.7	22.1–24.7	25.6–28.7	29.5–33.1	36.7–39.4	—	—	—	H42
24.8–27.0	24.8–27.0	28.8–33.0	33.2–35.1	39.5–41.4	—	—	—	H43
—	27.1–30.9	33.1–37.0	35.2–40.7	42.5–46.1	—	—	—	H44
—	31.0–35.5	37.1–43.1	40.8–44.3	46.2–50.3	45.0–49.5	—	—	H45
—	35.6–36.0	43.2–46.9	44.4–49.9	50.4–57.8	49.6–54.3	—	—	H46
—	—	—	—	57.9–60.6	54.4–58.4	—	—	H47
—	—	47.0–50.4	50.0–54.9	60.7–66.2	58.5–64.4	—	—	H48
—	—	50.5–57.1	55.0–63.3	66.3–75.4	64.5–70.0	—	—	H49
—	—	—	63.4–72.8	75.5–84.0	70.1–77.1	—	—	H50
—	—	57.2–60.0	72.9–79.3	84.1–90.2	77.2–85.5	—	—	H51
—	—	—	79.4–86.3	90.3–99.1	85.6–94.0	—	—	H52
—	—	—	86.4–90.0	—	94.1–104	—	—	H72
—	—	—	90.1–98.0	—	105–117	—	—	H73
—	—	—	—	99.2–108	—	—	—	H76
—	—	—	—	109–114	—	—	—	H77
—	—	—	—	115–129	—	—	—	H78
—	—	—	—	130–138	—	—	—	H79
—	—	—	98.1–108	—	118–130	—	—	H83
—	—	—	—	139–150	—	—	—	H85
—	—	—	—	151–165	—	—	—	H86
—	—	—	—	—	—	—	—	H89

FIGURE 13–9 Sample motor thermal overload selection table.

Many new motor starters are equipped with electronic overload relays. These overloads do not depend on the heat generated by the motor current passing through the heater element, but instead use a current sensor, such as a current transformer, to measure the current and provide a trip contact for the controller. This provides an advantage in that the overload can be adjusted to provide optimum protection without nuisance tripping over a wider range than that available in a thermal overload. The electronic overload can also be used to send operating data to a central location for remote monitoring. Other features include adjustable class of overload from 10 to 30, current imbalance protection, phase loss protection, phase reversal protection, and test function and trip indicators (see Figure 13–10).

(A)

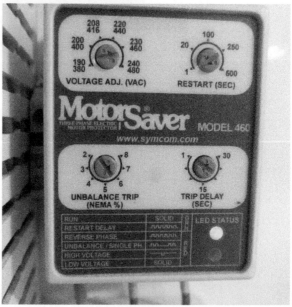

(B)

© Cengage Learning 2014

FIGURE 13–10 (A) Electronic overloads for motor controllers. (B) Voltage monitor to save motor from damage caused by voltage variations.

The class of overload refers to the speed and protection offered by the motor current monitor. Class 10-rated overloads are designed for hermetic refrigeration motors and for quick starting and fast trip on overload. (Trip time is less than 10 seconds, with six trips per hour.) Class 20 overloads are the most common and are designed for standard duty motors with normal time versus temperature curves. (Trip time is less than 20 seconds, with six trips per hour.) Class 30 overloads are used for long-time acceleration motors and extended high-current overloading. (Trip time is less than 30 seconds, with six trips per hour.) The trip currents are 125% of the minimum full-load current listed in the heater tables when the heater is at 40°C (104°F) ambient.

AUXILIARY CONTACTS

In addition to the standard contacts, a starter may be provided with externally attached auxiliary contacts, sometimes called *electrical interlocks* (Figure 13–11). These auxiliary contacts can be used in addition to the holding circuit contacts and the main or power contacts that

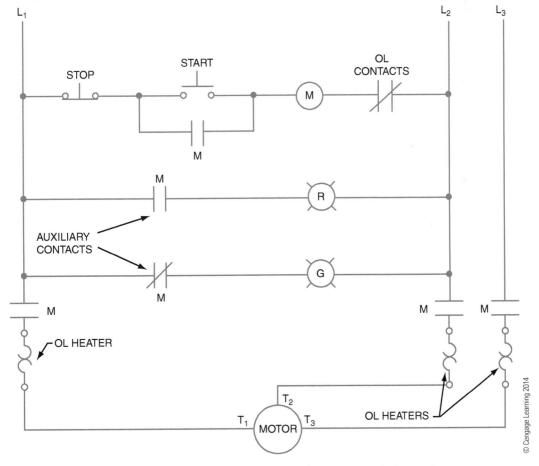

FIGURE 13–11 Electrical interlocks (auxiliary contacts) switch pilot lights in this circuit.

© Cengage Learning 2014

carry the motor current. Auxiliary contacts are rated to carry only control circuit currents of 0–15 amperes, not motor currents. Versions are available with either normally open (NO) or normally closed (NC) contacts, or combinations of NO and NC. Among a wide variety of applications, auxiliary contacts are used to

- control other magnetic devices where sequence operation is desired.
- electrically prevent another controller from becoming energized at the same time (such as reverse starting) called *interlocking*.
- make and break circuits of indicating or alarm devices, such as pilot lights, bells, or other signals.

Auxiliary contacts are packaged in kit form and can be added easily in the field.

ACROSS-THE-LINE MOTOR STARTER WITH REVERSING CAPABILITY

The direction of rotation of a squirrel-cage induction motor must be reversible for some industrial applications. To reverse the direction of rotation of three-phase motors, interchange any two of the three line lead connections to the motor terminals.

Figure 13–12 is an elementary wiring diagram of a motor starter with a reversing capability. When the three-power reverse contacts are closed, the phase sequence at the motor terminals is different from that when the three-power forward contacts are closed. Two of the line leads feeding to the motor are interchanged when the three reverse-power contacts close.

The control circuit has a pushbutton station with forward, reverse, and stop push buttons. The control circuit requires a mechanical and an electrical interlocking system provided by the pushbuttons. Electrical interlocking means that if one of the devices in the control circuit is energized, the circuit to a second device is open and cannot be closed until the first device is disconnected. Mechanical interlocks, shown by the broken lines in Figure 13–12, are used between the forward and reverse coils and pushbuttons.

Note in Figure 13–12 that when the forward pushbutton is pressed, it breaks contact with terminals 4 and 5, opening the reverse coil circuit, and makes contact between terminals 4 and 7. As a result, coil F is energized and the forward contacts close. The motor now rotates in the forward direction. If the reverse pushbutton is pressed, it breaks contact between terminals 7 and 8 and opens the circuit to coil F. This causes all forward contacts to open. As the reverse pushbutton is depressed further, it closes the contact between terminals 5 and 6 and energizes coil R. All reverse contacts are now closed, and the motor rotates in the reverse direction. If the stop button is pressed, the contact between terminals 3 and 4 is opened, the control circuit is interrupted, and the motor is disconnected from the three-phase source. The *National Electrical Code*® requirements for starting and running overload protection that apply to the ATL motor starter also apply to this type of motor starter.

Figures 13–12 and 13–13 are actually the same motor controller. Figure 13–12 is drawn in an elementary diagram. It has the control circuit in a schematic style, which shows the electrical relationship of the components. It shows the power contact of the magnetic starter below the

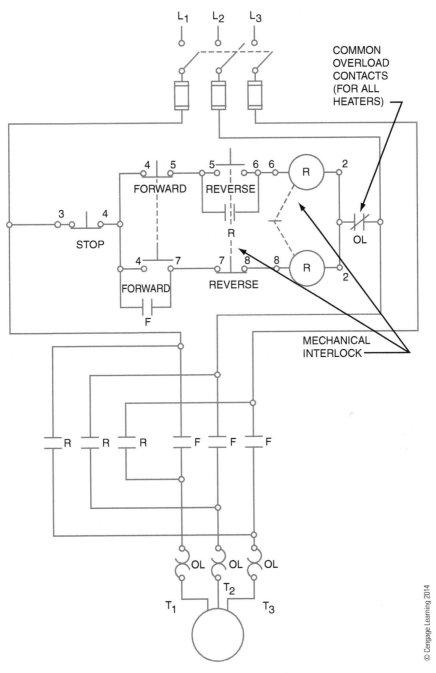

FIGURE 13–12 Elementary diagram of an ATL magnetic starter with reversing capabilities.

© Cengage Learning 2014

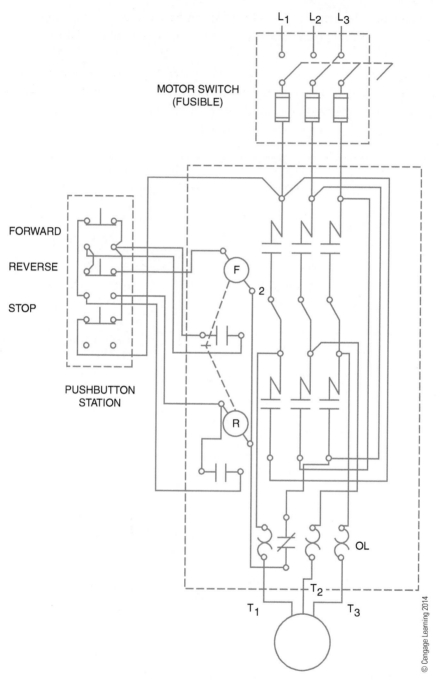

FIGURE 13–13 A panel, or wiring, diagram of an ATL magnetic starter with reversing capability.

schematic, and the electrical relationship of the motor control components. Figure 13–13 shows the same components, but in the approximate physical location of the components. This style of drawing is called a *wiring diagram*. Many electricians find it is easier to wire a panel from the wiring diagram as it shows physical location as well as general wire routing. Many electricians find it easier to troubleshoot from a schematic, or elementary, diagram, as it shows the electrical sequence of operation more clearly. It is important that you know how to read both types of drawings and be able to transfer from one to the other.

DRUM-REVERSING SWITCH

A drum-reversing switch, shown in Figure 13–14(A), can be used to reverse the direction of rotation of squirrel-cage induction motors.

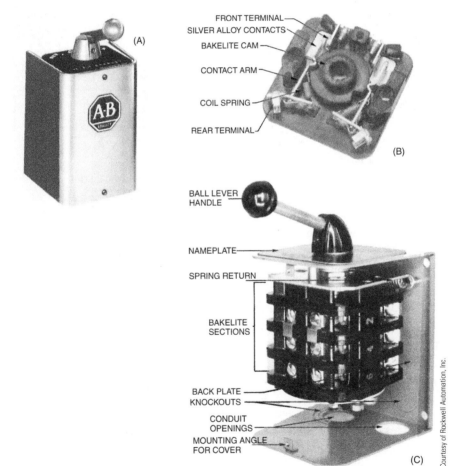

FIGURE 13–14 (A) Reversing-drum switch. (B) A bakelite section of a drum switch. (C) Bakelite section with cover removed.

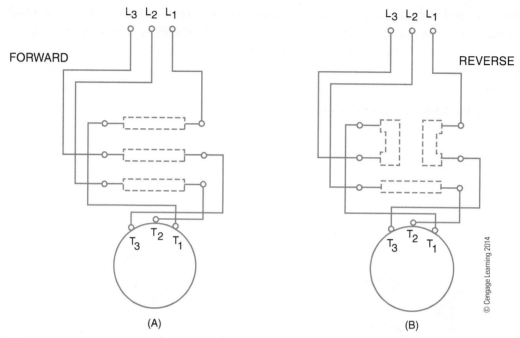

FIGURE 13-15 Connections for a drum-reversing switch: (A) forward; (B) reverse.

The motor is started in the forward direction by moving the handle of the drum-reversing switch from the off position to the forward (F) position. The connections for this drum controller in both the forward and reverse positions are shown in Figure 13–15. In the forward position, the switch connects line 1 to motor terminal 1, line 2 to motor terminal 2, and line 3 to motor terminal 3.

To reverse the direction of rotation, the drum switch handle is moved to the reverse (R) position. In the reverse position, line 2 is still connected to motor terminal 2. However, line 1 is now connected to motor terminal 3, and line 3 is connected to motor terminal T_1. When the handle of the drum switch is moved to the off position, all three line leads are disconnected from the motor.

SUMMARY

Many squirrel-cage motors are started with across-the-line motor starters. The motor and branch circuit should include short-circuit protection such as fuses or circuit breakers. The motor must also have running overload protection. This protection is usually found with the starter and is in the form of thermal-overload heaters or current-sensing overloads, and the associated overload relay. The overload relay is designed to open the control circuit to the motor starter in the event of a sustained overload on the motor. Motors can be automatically controlled by using a magnetic starter or may be manually controlled by using a drum type of controller. In either case, a three-phase motor can be reversed by interchanging two of the three line connections to the motor.

ACHIEVEMENT REVIEW

1. What is the purpose of starting protection for a three-phase motor?

_____

_____

_____

_____

2. What is the purpose of running overload protection for a three-phase motor?

_____

_____

_____

_____

3. What is meant by the code letter markings of squirrel-cage induction motors?

_____

_____

_____

_____

4. List some of the industrial applications for squirrel-cage induction motors with the code letter classification A.

_____

_____

_____

_____

5. List some of the industrial applications for squirrel-cage induction motors with the code letter classifications B to E.

_____

_____

_____

_____

6. List some of the industrial applications for squirrel-cage induction motors with the code letter classifications F to V.

_____

_____

_____

_____

7. A three-phase motor (code letter J) has a full-load current rating of 40 amperes, and a temperature rise of 40°C (104°F).

   a. What is the maximum size of fuses that can be used for branch-circuit protection?

   _____

   _____

   b. What size of heaters would be used for running overload protection?

   _____

   _____

8. What is the maximum starting protection allowed by the *National Electrical Code*®?

_____

_____

_____

_____

UNIT

14

CONTROLLERS FOR THREE-PHASE MOTORS

OBJECTIVES

After studying this unit, the student should be able to

- describe the basic sequence of actions of the following types of controllers when used to control three-phase AC induction motors:
  - jogging controller
  - quick-stop AC controller (plugging)
  - dynamic braking controller
  - resistance starter controller
  - automatic autotransformer compensator
  - automatic controller for wound-rotor induction motors
  - wye–delta controller
  - automatic controller for synchronous motors
- identify and use the various *National Electrical Code®* sections pertaining to controllers and remote-control circuits for motors.
- state why AC adjustable speed drives are used.
- list the types of adjustable speed drives.
- describe the operating principles of various AC adjustable speed drives.
- describe the basic controls used for medium-voltage motors.
- list the advantages and disadvantages of selected units.

The industrial electrician is required to install, maintain, and repair automatic AC controllers that start up and provide speed control for squirrel-cage induction motors, wound-rotor induction motors, and synchronous motors.

MOTOR CONTROLLERS WITH JOGGING CAPABILITY

Many industrial processes require that the driven machines involved in the process be inched or moved small distances. Motor controllers designed to provide control for this type of operation are called jogging controllers. *Jogging* is the quickly repeated closure of a controller circuit to start a motor from rest for the purpose of accomplishing small movements of the driven machine.

An across-the-line (ATL) magnetic motor switch may be used to provide jogging control if the proper type of pushbutton station is used in the control circuit. Such a pushbutton station is called a *start-jog-stop station.*

Figure 14–1(A) is a diagram of the connections for a three-phase AC induction motor connected to a jogging, ATL motor-starting switch.

Figure 14–1(B) shows the starter with the cover removed. Note in Figure 14–1 that the connections and operation of the start and stop pushbuttons are the same as those for a standard pushbutton station with start and stop positions. The connections for the jog pushbutton, however, are more complex and should be studied in detail. When the jog pushbutton is pressed, coil M is energized, main contacts M close, and the motor starts turning. The small auxiliary contacts M also close, but do not function as a sealing circuit around the jog pushbutton because pushing the jog pushbutton also opens the sealing circuit. As a result, as soon as the jog pushbutton is released, coil M is de-energized and all M contacts open. Before the jog pushbutton returns to its normal position, the sealing M contacts open and thus the control circuit remains open. This control also can be used for standard start–stop operations. In summary, then, repeated closures of the jog pushbutton start the motor momentarily so that the driven machine can be inched or jogged to the desired position.

QUICK-STOP AC CONTROLLER (PLUGGING)

Some industrial applications require that three-phase motors be stopped quickly. If any two of the line leads feeding a three-phase motor are reversed, a countertorque is set up that brings the motor to a quick standstill before it begins to rotate in the reverse direction. If the circuit is interrupted at the instant the motor begins to turn in the opposite direction, the rotor just stops. This method of bringing a motor to a quick stop is called *plugging.* The motor controller required to provide this type of operation is an ATL magnetic motor starter with reversing control and a special plugging relay. The plugging relay may be belt driven from an auxiliary pulley on the motor shaft, or attached to a through-shaft motor.

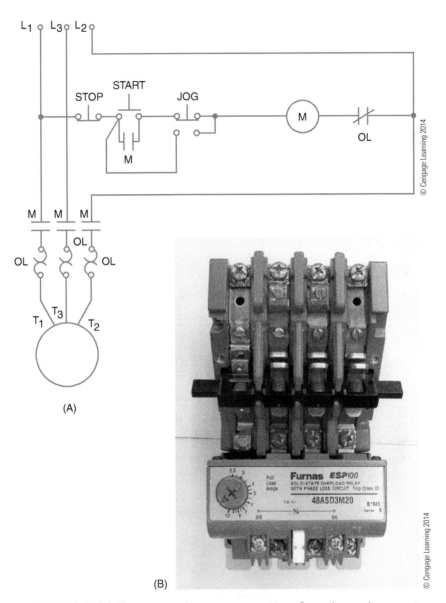

FIGURE 14–1 (A) Elementary diagram connections for a three-phase motor with jogging. (B) Magnetic starter with cover removed-exposing contacts.

The connections for a quick-stop AC controller are shown in Figure 14–2(A). The controller itself is shown in Figure 14–2(B). When the start pushbutton is pressed, relay coil F is energized. As a result, the small, normally closed F contacts open. These contacts are connected in series with the reverse coil, which locks out reverse operation. In addition, the other small,

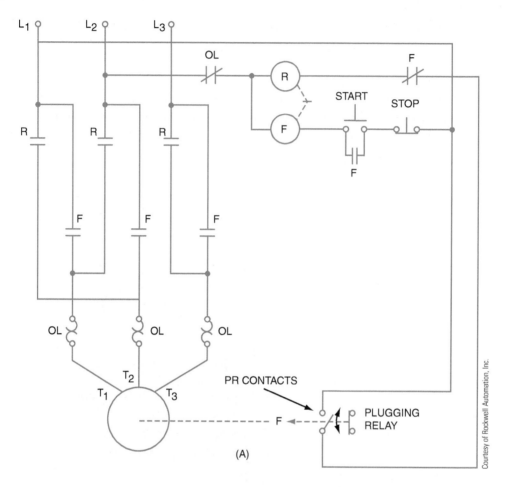

(A)

(B)

Courtesy of Rockwell Automation, Inc.

Courtesy of Rockwell Automation, Inc.

FIGURE 14-2 (A) Elementary circuit with a plugging relay. (B) AC full-voltage reversing starter, size 1.

normally open F contacts close and maintain the start pushbutton circuit. When the start button is released, the circuit of coil F is maintained through the sealing circuit, main F contacts close, and the rated three-phase voltage is applied to the motor terminals. Then the motor comes up to speed, and PR contacts of the plugging relay close.

When the stop pushbutton is pressed, the F relay coil is de-energized. As a result of this action, the main F contacts for the motor circuit open and disconnect the motor from the three-phase source. In addition, the small F holding contacts open, which de-energizes the holding or sealing circuit around the start pushbutton. Finally, the small F contacts in series with the reverse relay close to their normal position, and the reverse relay coil is energized.

The main R contacts now close to reconnect the three-phase line voltage to the motor terminals. The connections of two line leads are interchanged. The resulting reversing countertorque developed in the motor brings it to a quick stop. At the moment the motor begins to turn in the reverse direction, PR contacts open due to the mechanical action of the PR relay unit. Coil R is de-energized and the R contacts open and interrupt the power supply to the motor. Because the motor is just beginning to turn in the opposite direction, it comes to a standstill. The motor supplies the mechanical power to drive a disk that causes PR contacts to close when the motor is in operation.

DYNAMIC BRAKING WITH INDUCTION MOTORS

From the study of DC controllers, we know that dynamic braking is a method used to help bring a motor to a quicker stop without the extensive use of friction brakes. In this application, dynamic braking means that the motor involved is used as a generator. An energy-dissipating resistance is connected across the terminals of the motor after it is disconnected from the line.

Dynamic braking also can be applied to induction motors. When the stop pushbutton is pressed, the motor is disconnected from the three-phase source, and the stator windings are excited by a DC source. A stationary magnetic field is developed by the direct current in the stator windings. As the squirrel-cage rotor revolves through this stationary field, a high rotor current is created. This rotor current reacts with the stationary field of the stator to produce a countertorque that slows and stops the motor.

Figure 14–3 is a diagram of an AC motor installation with an across-the-line magnetic motor starter and dynamic braking capability.

When the start pushbutton is pressed, coil M is energized. At this instant, the main M contacts close and connect the motor terminals to the three-phase source, and the auxiliary, normally open M contacts close and provide a maintaining circuit around the start pushbutton. When relay coil M is energized, the normally closed contacts M in the DC control circuit open, with the result that both the main DC relay coil N and the time-delay relay coil TR are de-energized, and interlocks in the DC circuit open. The three-phase voltage applied to the motor terminals causes the motor to accelerate to the rated speed.

When the stop pushbutton is pressed, coil M is de-energized. At this moment, a number of actions occur: (1) The main contacts M open and disconnect the motor from the

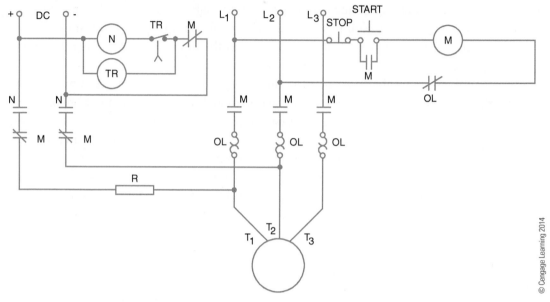

FIGURE 14–3 A circuit for dynamic braking of an AC motor.

three-phase source; (2) the auxiliary M contacts open (these contacts act as a maintaining circuit); (3) M protective interlocks in the DC circuit close; and (4) the auxiliary, normally closed M contacts in the DC control circuit close and energize the time-delay relay and the main relay coil N. Energizing relay coil N causes the closing of the N contacts so that DC voltage is connected on the AC windings through a current-limiting resistance. As a result, the motor comes to a quick stop. Following a definite period after the motor has stopped, measured in seconds, the TR relay coil operates and opens the TR contacts to cause coil N to become de-energized. Thus, the N contacts open and disconnect the motor windings from the DC source. The controller now is ready for the next starting cycle.

Timing contacts are shown in their de-energized condition. Timers are either *on delay* or *off delay* and are used in motor control work. The actual timer mechanism varies depending on the vintage of the controller and the manufacturer. See Unit 10 for a complete description of the timer symbols and operations.

RESISTANCE STARTER CONTROLLER

When a squirrel-cage induction motor is connected directly across the rated line voltage, the starting current may be 300% to 600% or more of the rated current of the motor. In large motors, this high current may cause serious voltage regulation problems and overload industrial power feeders.

The starting current of a squirrel-cage induction motor can be reduced by using a resistance starter controller. This type of controller inserts equal resistance values in each line wire at the instant the motor is started. After the motor accelerates to a value near its rated speed, the resistance is cut out of the circuit and full line voltage is applied to the motor terminals.

Figure 14–4(A) shows a diagram of the circuit connections for a resistance starter. A photo of this starter is shown in Figure 14–4(B). When the start button is pressed, main relay coil M is energized. The main contacts close and connect the motor to the three-phase source through the three resistors (R). The circuit for coil M is maintained through the small auxiliary contacts (3 and 4), which act as a sealing circuit around the normally open start pushbutton. When the main contacts of relay coil M are closed, a timing relay is energized. The on-delay timer closes contacts 4 to 5 after a set time. After a predetermined time elapses, the definite time contacts close and energize coil R. Coil R causes three sets of contacts to close and shunt out the three resistors. Thus, the motor is connected directly across the rated line voltage with no interruption of the power line (closed transition).

When the stop button is pressed, the circuits of both coil M and coil R are opened. This causes the opening of the main contacts, the sealing contacts, and the contacts that shunt the series resistors. As a result, the motor is disconnected from the three-phase source.

The motor starting current in the resistance starter causes a relatively high voltage drop in the three resistors. Because of this, the voltage across the motor terminals at starting is low. As the motor accelerates, the current decreases; the voltage drop across the three resistors decreases; and the terminal voltage of the motor increases gradually. A smooth acceleration is obtained because of this gradual increase in the terminal voltage. However, it may be unwise to select resistance starting for many starting tasks because of the energy dissipated in the starting resistors.

AUTOMATIC AUTOTRANSFORMER COMPENSATOR

An automatic compensator is used to automatically connect an autotransformer in series with the motor during the starting cycle (Figure 14–5). A *compensator* is another name for an autotransformer starter. By inserting an autotransformer, the voltage to the motor is reduced, and therefore the starting current is reduced. The starting current drawn from the line is reduced. Because the autotransformer steps down the voltage to the motor, the secondary (motor) current is higher than the primary current. The result is that the current drawn from the line is much less than if the motor were started with a reduced voltage through a resistor. The automatic compensator can be pushbutton controlled from a convenient location. Figure 14–5(B) shows a typical schematic wiring diagram for an automatic autotransformer compensator.

When the start button is pressed, a circuit is established from power line 1 through the following devices to power line 2: through the stop button and the depressed start button to

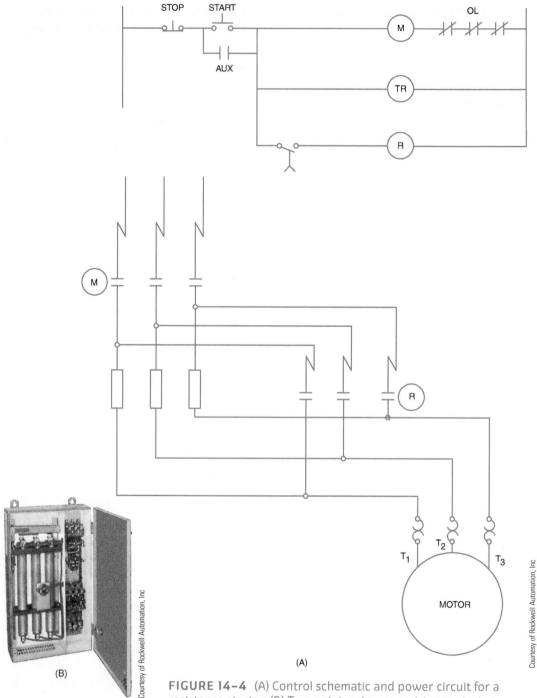

FIGURE 14–4 (A) Control schematic and power circuit for a resistance starter. (B) Two-point, primary-resistance starter rated at 25 hp, 600 volts.

Courtesy of Rockwell Automation, Inc

Courtesy of Rockwell Automation, Inc

(A)

(B)

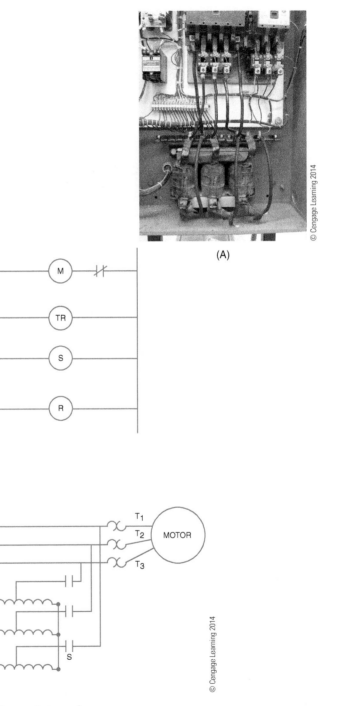

(A)

FIGURE 14-5 (A) Reduced voltage starter, autotransformer type. (B) Control schematic and power circuit for autotransformer type of starter.

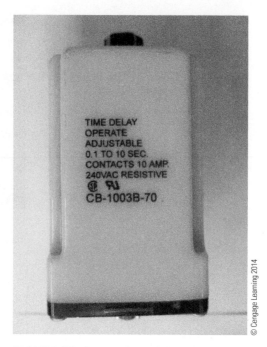

TIME DELAY
OPERATE
ADJUSTABLE
0.1 TO 10 SEC.
CONTACTS 10 AMP.
240VAC RESISTIVE
CB-1003B-70

© Cengage Learning 2014

FIGURE 14–6 An AC on-delay timer that provides time delay after the coil is energized.

energize the M coil on control line 1 and the timing relay coil on control line 3 and through the TR timer contacts on line 4 to the S coil. The M auxiliary coil on line 2 closes and provides a holding circuit for the start button. With the M coil energized, the M power contacts close and the S power contacts close. This supplies power to the autotransformer, and a reduced voltage is supplied to the motor. As the TR timer times out, the TR contacts on line 5 close and the TR contacts on line 4 open. As the contacts on line 4 open, the S relay is de-energized and the S contacts on line 5 close to supply control power to the R relay. Now the power is disconnected from the autotransformer and supplied directly to the motor as full voltage (see Figure 14–6).

AUTOMATIC CONTROLLER FOR WOUND-ROTOR INDUCTION MOTORS

Manual speed controllers, such as the faceplate type or the drum type, may be used to provide speed control for wound-rotor induction motors in industrial applications. If the resistance in the rotor circuit of a wound-rotor induction motor is to be used only on starting, then an automatic controller may be used (Figure 14–7). In this case, resistors in the rotor circuit are automatically removed by contactors arranged to operate in sequence at definite time intervals.

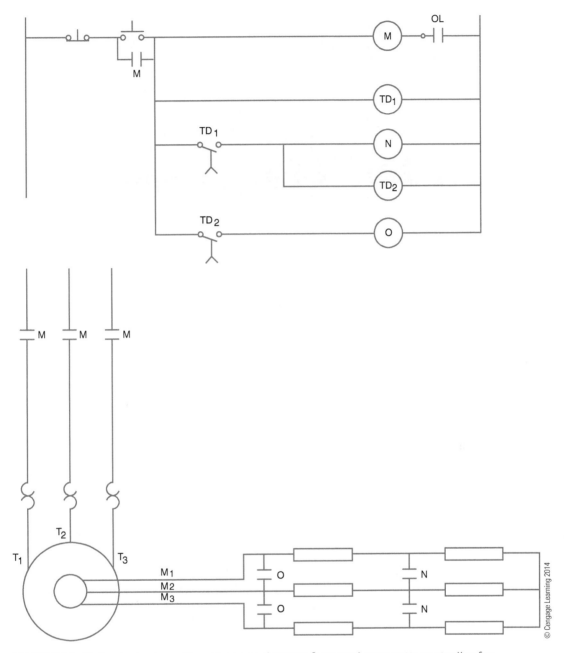

FIGURE 14–7 Control schematic and power diagram for timed automatic controller for wound-rotor motor.

As shown in Figure 14–7, when the start button is pressed, the main relay coil M is energized. The main contacts are closed to connect the stator circuit of the motor directly across the three-phase line voltage. All the resistance of the controller is inserted in the secondary circuit of the motor as it begins to accelerate.

Courtesy of Schneider Electric USA, Inc.

FIGURE 14–8 Solid-state timing relays with various plug-in program keys.

When the start pushbutton is released, the M coil auxiliary contact holds the M coil in, and power is applied to the motor primary windings marked T_1, T_2, and T_3. The first timer begins timing at the same instant the M coil is energized. After a predetermined time, timer 1 times out and closes the contacts on the next line, thereby energizing the N coil. This pulls the N power contacts closed and shunts some of the resistance in the rotor secondary. As the N coil is energized, the second timer, TD_2, is started. After another predetermined time, TD_2 times out and closes, energizes the O coil, and pulls the remaining power contacts closed to shunt all the secondary rotor resistance. Examples of timing relays appear in Figure 14–8 and Figure 14–9.

When the stop pushbutton is pressed, relay coil M is de-energized, and contacts M open to disconnect the motor from the line. Coil O also is de-energized, and contacts O open, with the result that all resistance is inserted in the rotor circuit for the next starting cycle.

Wound-rotor motors are also controlled by electronic controllers. Often these motors used electromechanical controllers to place resistance into the secondary rotor circuit to control inrush current and then to control speed. With automatic controllers, these motors can be retrofitted to use electronic soft-start control to bring the motor up to speed with a minimum inrush current. If additional speed control is needed, the controller still can insert power resistors into the secondary rotor circuit. The more resistance inserted, the slower the speed of the motor.

National Electrical Code® regulations for wire size, starting overload protection, and running overload protection apply to both manual speed controllers and automatic controllers used with wound-rotor induction motors.

FIGURE 14-9 Small solid state timers can provide many functions such as "on-delay".

WYE–DELTA CONTROLLER

Figure 14–10(A) shows a simple method by which a three-phase, delta-connected motor can be started on reduced voltage by connecting the stator windings of the motor in wye during the starting period. Figure 14–10(B) shows the actual starter. After the motor accelerates, the windings are reconnected in delta and placed directly across the rated three-phase voltage.

When the Start button is pressed, the main M contacts close, and relay coil Y and time-delay relay TR are energized. Coil Y causes contacts Y to close, and the windings of the motor are connected in wye. If the line voltage is 230 volts, the voltage across each winding is

$$\frac{230}{1.73} = 133 \text{ volts}$$

The voltage across each winding is only 58% of the line voltage when the windings are connected in wye at the start position. (See three-phase voltage in *Electricity 3*, Unit 8.)

After a definite period of time, the time-delay relay TR opens the circuit of relay coil Y, the Y contacts open, and the Y interlocks close.

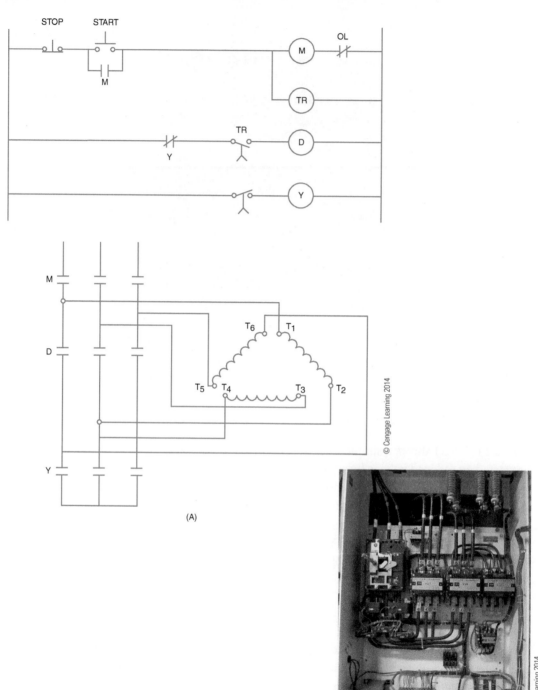

FIGURE 14–10 (A) Control schematic and power diagram for wye–delta starter with open transition. (B) Photo of a wye–delta, 200 hp closed transition.

(A)

(B)

The time-delay relay TR then closes the circuit of relay coil D (delta). All D contacts are closed, and the motor winding connections are changed from wye to delta. Full line voltage is applied to the motor windings, and the motor operates at its rated capacity.

Motors started by a wye–delta controller must have the leads of each phase winding brought out to the terminal connection box of the motor. In addition, the phase windings must be connected in delta for the normal running position.

Note: The electrician should never attempt to operate a three-phase, wye-connected motor with this type of controller. This results in an excessive voltage applied to the motor windings in the run position when the windings are connected in delta by the controller.

Another method of controlling the wye–delta controller is to provide a way to change from the wye connection to the delta connection without completely disconnecting the motor from the power line. This method is called *closed transition* as compared to the previously described *open transition*. By using resistors in the power circuit, the transition can be made from wye to delta, as shown in Figure 14–11. When the start button is pressed, the S coil on line 1 and the 1M coil on line 2 are energized, and the power is supplied to the motor, which is connected in the star pattern by the S power contacts. As the timer 1M closes, it energizes coil 1A on line 3 and resistors are connected into the power circuit of the motor for a brief time, and the S coil is de-energized, releasing the motor from the wye connection. As 1A timing contacts close on line 4, the 2M coil is energized and the resistors are effectively shunted out of the circuit.

When using closed transition, the small current fluctuation that occurs as the motor is briefly disconnected from the line may cause a voltage dip in the voltage supply as well as a fluctuation in the motor's developing torque. To reduce the voltage and current disturbance, closed transition is used. Note the fluctuations in Figure 14–12.

COMPARISON OF REDUCED-VOLTAGE STARTING METHODS

The chart in Figure 14–13, shows the advantages and limitations of the various methods of reduced-voltage/reduced-torque types of motor starting. The advantage of using the reduced voltage is to reduce the inrush current to the motor from the supply line, thereby reducing the voltage dips and disturbances to the facility's electrical system. The motor's normal starting torque is also reduced, which may be an advantage or disadvantage. The starting torque is reduced in proportion to the voltage reduction squared. If the voltage is reduced to 50%, then the torque is 0.5^2, or $0.25 = 25\%$. If the voltage is reduced to 80%, the torque is reduced to 64%. An advantage to reduced torque is that the driven machinery is not violently started from a stopped position to full-locked rotor torque. The disadvantage may be that there is not enough torque to start the load.

AUTOMATIC CONTROLLER FOR SYNCHRONOUS MOTORS

Synchronous motors may be started by an ATL magnetic motor starting switch, a manual starting compensator, or an automatic starting compensator. Dynamic braking may be provided by the controller.

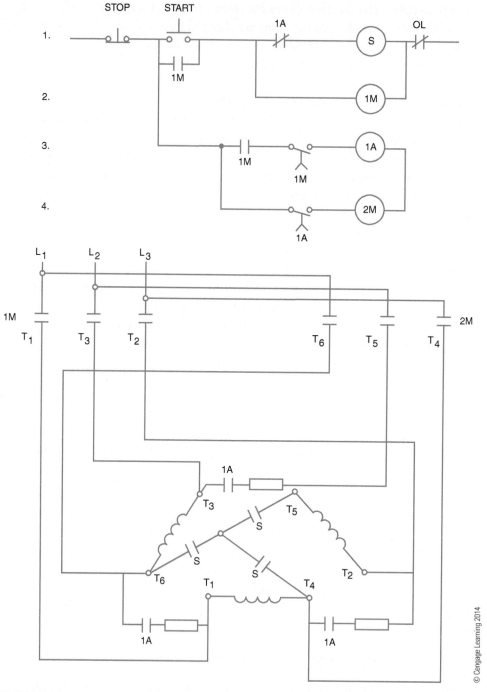

FIGURE 14–11 Control schematic and power diagram for a wye–delta starter with a closed transition.

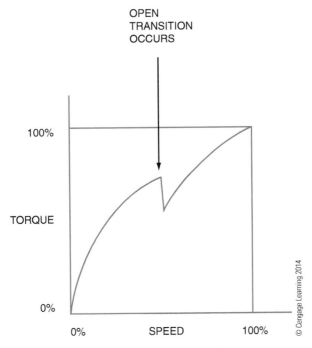

FIGURE 14-12 Open transition speed torque curve shows a dip when transition occurs.

Figure 14–14 is a diagram of the connections for a synchronous motor controller with dynamic braking. When the start button is pressed, main relay coil M is energized. The four normally open M contacts close, and the two normally closed M contacts open. Three-phase voltage is applied to the motor terminals. When the motor accelerates to a speed near the synchronous speed, the DC field circuit is energized by secondary controls.

When the stop button is pressed, main relay coil M is de-energized. The M contacts open and disconnect the motor terminals from the three-phase line. The two normally closed M contacts reconnect the motor windings through the resistors, and the DC field remains energized. As a result, the synchronous motor acts as an AC generator and delivers electrical energy to the two R resistors. The use of this type of controller results in a more rapid slowing of a synchronous motor.

The *National Electrical Code*® provides guidelines for branch-circuit fuse protection and running overload protection for branch circuits feeding three-phase synchronous motors, and for allowable conductor sizes for branch-circuit–feeding synchronous motors. Local building and electrical code authorities should be consulted before installations are made with motors and controllers that do not comply with *National Electrical Code*® rulings.

Type of Starter	Starting Characteristics			Advantages	Limitations
	Voltage at Motor	Line Current	Starting Torque		
Full voltage	100%	100%	100%	• Lowest cost. • Less maintenance. • Highest starting torque.	• Starting inrush current may exceed limits of electrical distribution system. • Starting torque may be too high for the application.
Auto-transformer	80 65 50	64 42 25	64 42 25	• Provides most torque per ampere of the line current. • Taps on autotransformer permit adjustment of starting voltage. • Suitable for long starting periods. • Closed transition starting.	• In lower hp ratings is most expensive design. • Heavy, physically largest type. • Low power factor. • Most complex of reduced voltage starters because proper sequencing of energization must be maintained.
Primary resistance	65	65	42	• Least complex method to obtain reduced voltage starting characteristics on low-capacity systems because interlocking of contactors is unnecessary. • Smoothest acceleration of electromechanical types. • Improves starting power factor because voltage current lag is shortened by putting a resistance in series with the motor. • Less expensive than auto-transformer starter in lower hp ratings.	• Additional power loss in resistors compared with other types of starters. • Low torque efficiency (decreases as voltage is decreased). • Starting characteristics not easily adjusted after manufacture. • Duty cycle may be limited by resistor rating. • High initial inrush current.
Part winding	100	65	42	• Starter less expensive than other types of reduced voltage control. • Closed-circuit transition. • Most dual-voltage motors can be started part winding on lower of two voltages. • Control smaller than other types.	• Torque efficiency usually poor for 3600 RPM motors. • Possibility of motor not fully accelerating due to torque dips. • Unsuitable for high-inertia, long-standing loads. • Requires special motor design for voltages other than 230 volts.
Wye delta	100	33	33	• Low torque efficiency. • No torque dips or unusual winding stresses occur as in part winding starting.	• Requires special motor design. • Starting torque is low. • Usually not suitable for high-inertia loads. • Control more complex than many other starter types.
Solid state	Adjust	Adjust	Adjust	• Includes constant current, ramped current, or tachometer type starting. • Adjustable current limit and starting time. • Increased duty cycle compared with electromechanical types. • Power factor controller and line voltage limiting included. • Multiple adjustable points over wide range. • Smoothest acceleration.	• Specialized maintenance required. • Shorting contactor is required for NEMA 4 and 12 enclosure. • Ventilation required. • Higher priced. • Isolation contactor may be required.

FIGURE 14–13 Comparison chart for reduced-voltage starters.

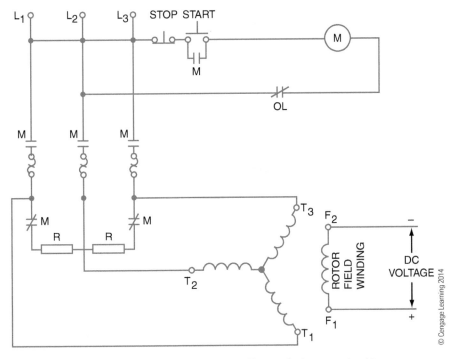

FIGURE 14–14 Synchronous motor controller with dynamic braking.

SOLID-STATE REDUCED VOLTAGE STARTERS

Solid-state devices and equipment are used for reduced-voltage motor starting, electrical-energy–saving control circuits, variable speed drives, motor protection, and other applications. A motor starter consists of a control circuit, a motor power circuit, and protective devices for the wiring and the motor. The functions of a starter are performed by contactors and overload relays in electromechanical motor starters. In solid-state starters, the control functions are performed by semiconductors. They are controlled by integrated circuits and microprocessors to provide the protective functions, operating instructions, and control.

Construction and Operation

The solid-state, reduced-voltage starter provides a smooth, stepless acceleration of a three-phase induction motor. This is accomplished by gradually turning on six power silicon-controlled rectifiers (SCRs). Two SCRs per phase are connected in a back-to-back or reverse parallel arrangement (Figure 14–15). The SCRs are mounted on a heat (dissipating) sink to make up a power pole (phase). The SCRs are connected back to back so that they can pass AC and control the amount of voltage. Each power pole contains the gate firing circuits as discussed in Unit 11.

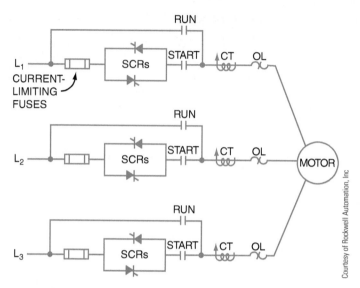

FIGURE 14–15 Solid-state, reduced-voltage starter power circuit.

An integrated thermal sensor is also provided to de-energize the starter if an over-temperature condition exists.

The firing circuitry on each power pole is controlled by a logic module. These modules monitor the starter for correct start-up and operating conditions. Some motor starters provide a visual indicator of the starting condition through the use of light-emitting diodes (LEDs).

The current-limiting starter is a common type that is designed to maintain the motor current at a constant level throughout the acceleration period. A current-limit potentiometer adjustment is provided to preset this current. A starter with current ramp acceleration is designed to begin acceleration at a low current level and then increase the current during the acceleration period.

As indicated in Figure 14–15, this starter includes both *start* and *run* contactors. The start contacts are in series with the SCRs; the run contacts are in parallel with the combination of SCRs and start contacts. When the starter is energized, the start contacts close. The motor acceleration is then controlled by phasing on the SCRs. When the motor reaches full speed, the run contacts close, and the motor is connected directly across the lines (closed transition). At this point, the SCRs are turned off and the start contacts open. Under full-speed running conditions, the SCRs are out of the circuit, eliminating SCR power losses during the run cycle. This feature saves energy; it also guards against possible damage due to overvoltage transients. With the starter in the de-energized position, all contacts are open, isolating the motor from the line. This open-circuit condition

Courtesy of Rockwell Automation, Inc

FIGURE 14–16 Solid-state, reduced-voltage starter power circuit.

protects against accidental motor rotation as a result of SCR misfiring and/or SCR damage caused by overvoltage transients. A solid-state, reduced-voltage starter is shown in Figure 14–16. Field connections are very similar to those for electromechanical starters.

Reduced Voltage Operation

To reduce the voltage applied to the motor in a solid-state starter, the SCRs can be turned on by the "gate" electrode for any desired part of each half cycle. Usually the SCRs turn off as the current wave reaches zero. They stay off until gated on again in the next half cycle. Some devices can vary the switching and timing. By switching the controlled current gating, the effective AC voltage can be varied to the motor. This voltage can be varied from zero to full voltage as required. The voltage is applied at some preset minimum value that can start the motor rotating. As the motor speed builds up, the SCR "on" time is gradually increased. The voltage is increased until the motor is placed across the line at full voltage. Mechanical shock is reduced, and the current inrush can be regulated and controlled as desired (Figure 14–17). The solid-state, reduced-voltage starter can replace any of the electromechanical starters already described for reduced-voltage starting.

CODE *REFERENCES FOR MOTOR CONTROLLERS*

The following sections of the *National Electrical Code®* are concerned with motor controllers and remote control circuits.

1. *430.8* and *430.9* refer to the identification of motors and controllers with respect to controller nameplate ratings and terminal markings.

2. *430, Part III*, is concerned with overload protection.

3. *430.37* provides the number of running overcurrent relays required for various electrical systems.

4. *430, Part IV*, is concerned with branch-circuit, short-circuit, and ground-fault protection.

5. *430, Part VI*, is concerned with controller circuits.

6. *430.81, Part VII*, is concerned with motor controller installations.

7. *430.101, Part IX*, covers motor disconnecting means.

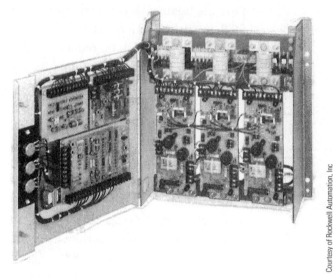

Courtesy of Rockwell Automation, Inc

FIGURE 14–17 SCR controller section, the regulating part of the starter. The controller determines to what degree the SCRs should be phased on, thereby controlling the voltage applied to the motor.

AC ADJUSTABLE SPEED DRIVES

Adjustable speed drives have a flexibility that is particularly useful in specialized applications. For this reason, these drives are widely used throughout the industry for conveyors used to move materials, hoists, grinders, mixers, pumps, variable speed fans, saws, and crushers. The advantages of AC drives include the maximum use of the driven equipment, better coordination of production processes, and reduced wear on mechanical equipment.

The AC induction motor is the major converter of electrical energy into another usable form. About two-thirds of the electrical energy produced in the United States is delivered to motors.

Much of the power that is consumed by AC motors goes into the operation of fans, blowers, and pumps. It has been estimated that approximately 50% of the motors in use are for these types of loads. Such loads are particularly appropriate to consider for energy savings. Several alternate methods of control for fans and pumps have been developed and show energy savings over traditional methods of control.

Fans and pumps are designed to meet the maximum demand of the system in which they are installed. Often, however, the actual demand varies and may be much less than the design capacity. Such conditions are accommodated by adding outlet dampers to fans or throttling valves to pumps. These controls are effective and simple but affect the efficiency of the system.

Other forms of control have been developed to adapt machinery to varying demands. These controls do not decrease the efficiency of the system as much as the traditional methods of control. One of the newer methods is the direct, variable speed control of the fan or pump. This method produces a more efficient means of flow control compared to other existing methods.

In addition to a tangible reduction in the power required to operate equipment and machinery resulting from the use of adjustable speed drives, other benefits include extended bearing life and pump seal life.

TYPES OF ADJUSTABLE SPEED DRIVES

Several types of variable speed drives can be used with wound-rotor induction motors. These drives are eddy-current (magnetic) drives, variable pitch drives, and adjustable-frequency drives.

Eddy-Current (Magnet) Drives

The eddy-current drive couples the motor to the load magnetically (Figure 14–18). The electromagnetic coupling is a simple way to obtain an adjustable output speed from the constant input speed of squirrel-cage motors. No mechanical contact exists between the rotating members of the eddy-current drive; thus, there is no wear. Torque is transmitted between the two rotating units by an electromagnetic reaction created by an energized rotating coil winding. The rotation of the ring with relation to the electromagnet generates eddy currents and magnetic fields in the ring. Magnetic interaction between the two units transmits torque from the motor to the load. The slip between the motor and the load can be controlled continuously with great precision.

Torque can be controlled using a thyristor in an AC or DC circuit, or by using a rheostat to control the field through slip rings. When the eddy-current drive responds to an input or command voltage, the speed of the driven machine changes. A further refinement can be obtained in automatic control to regulate and maintain the output speed. The magnetic drive can be used with nearly any type of actuating device or transducer that can provide an electrical signal. For example, the input can be provided by static controls and sensors that detect liquid level, air and fluid pressures, temperature, and frequency.

Magnetic eddy-current drives are used for applications requiring an adjustable speed such as cranes, hoists, fans, compressors, and pumps (Figure 14–19).

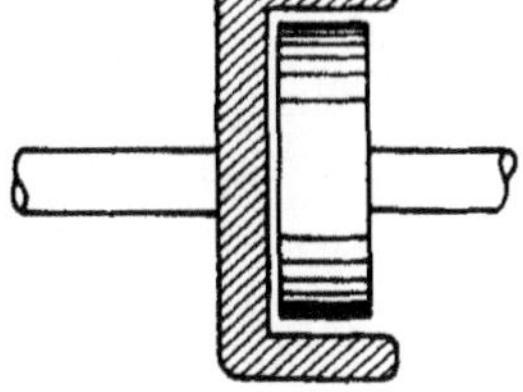

© Cengage Learning 2014

FIGURE 14–18
Spider-rotor coil magnet rotated within a steel drum.

Variable Pitch Drives

The speed of an AC squirrel-cage induction motor depends on the frequency (hertz) of the supply current and the number of poles of the motor. The equation expressing this relationship is

FIGURE 14-19 Two magnetic drives driven by 100 hp induction motors mounted on top.

$$RPM = \frac{60 \times Hertz}{Pairs\ of\ poles}$$

or

$$Synchronous\ speed\ (RPM) = \frac{120 \times Hertz}{Number\ of\ poles}$$

A frequency changer may be used to vary the speed of this type of motor. A possible method is to drive an alternator through an adjustable mechanical speed drive.

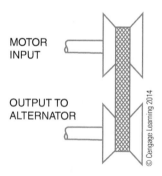

FIGURE 14-20 Variable pitch pulley method of obtaining continuously adjustable speed from constant motor speed shaft.

The voltage is regulated automatically during frequency changes. An AC motor drives a variable cone pulley or sheave, which is belted to another variable pulley on the output shaft (Figure 14–20). When the relative diameters of the two pulleys are changed, the speed between the input and the output can be controlled. As the alternator speed is varied, the output frequency varies, thereby varying the speed of the motor, or motors, connected electrically to the alternator supply.

Adjustable Frequency AC Drives

Adjustable frequency (static solid-state) drives are also commonly called variable frequency drives (VFDs). The power conversion losses are greatly reduced when using these transistor controllers for adjustable speed drives. They are available in a

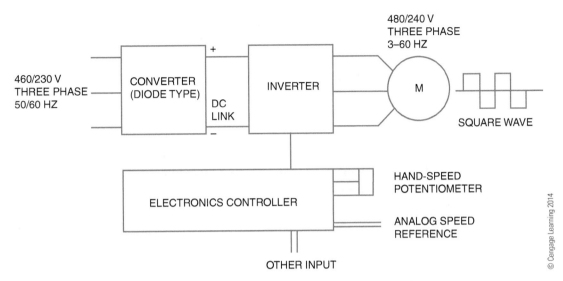

FIGURE 14-21 Controller operation.

range of horsepowers from fractional to 1000 hp. Adjustable frequency drives are designed to operate standard AC induction motors although "inverter duty" motors are more reliable when using a variable frequency input. This allows them to be added easily to an existing system (Figure 14–21).

Where energy saving is a major concern, the drives are ideal for pumping and fan applications. They are also used for many process control or machine applications where performance is a major concern. Many adjustable speed precision applications were limited to the use of DC motors. By using adjustable frequency controllers with optional dynamic braking, standard squirrel-cage motors can now be used in these applications. Municipal, industrial, commercial, and mining applications include sewage, waste water, slurry and booster pumps, ventilation and variable air volume fans, conveyors, production machines, and compressors.

AC VARIABLE SPEED/VARIABLE FREQUENCY CONTROLS

AC motors are designed to run at a specific full-load speed. This design speed takes into account various losses in the motor, including copper losses of the stator and the rotor and other losses such as iron losses, friction, and windage. The end result is some speed less than synchronous speed. Synchronous speed is calculated by the following formula:

$$\text{Synchronous RPM} = \frac{120 \times \text{Frequency}}{\text{Number of poles}}$$

Use this formula to determine the synchronous speed of a motor if the number of poles and the applied frequency are known. The number of poles is usually fixed, and the frequency of a normal power feed in the United States is 60 Hz. Many other countries of the world use 50 Hz. as their standard frequency. Therefore, to operate an AC motor at other than its design speed, either the number of poles or the frequency must be altered. Some adjustable speed motors are able to reconnect the poles or connect a separate winding of poles to establish other set speeds. To vary the speed of an AC motor over a wide range of speeds, the applied frequency is altered.

There are two basic techniques for altering the frequency of the applied power through electronic means. Both techniques use the principle of rectifying the three-phase, AC, 60 Hz input power to a DC supply (see Figure 14–22). The DC is then filtered to provide smooth DC to the inverter section of the electronic controller. The amplitude of the output voltage must change with the frequency, because at low frequency the impedance of the motor is low and the voltage must also be reduced to prevent overheating of the motor. Conversely, as the frequency is raised above 60 Hz, the motor's impedance is increased and voltage must also increase to maintain motor torque. This parameter of the controller functions is known as the volts per hertz ratio.

One method of speed control is the variable voltage inverter. The DC voltage applied to the inverter is adjusted, and then the pulse is modified to create various frequencies (see Figure 14–21).

The other and more common method is known as *pulse width modulation (PWM)*. The voltage output of the DC-to-AC inverter is really a series of DC pulses that is stepped to produce a "staircase" approximation of the sinewave at the frequency desired to control the motor speed. The controller uses a sensor and a set point to determine output frequency and the desired set speed.

The control modules adjust the output of a DC voltage-control module as it adjusts the output of the frequency-control module. The frequency-control module typically drives the output power controller. There are two output power controllers for each of the three phases: one controls the positive half cycles of one phase, and the other controls the negative half cycles of the same phase. Six output modules are required for three phases (Figure 14–23).

When connected to a motor, the stepped output waveform appears close to a sinewave because of the motor's inductance. Frequency then determines the speed of the motor in conjunction with the set number of poles.

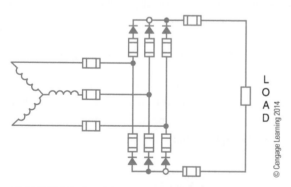

FIGURE 14-22 Three-phase, full-wave rectifier with connected load.

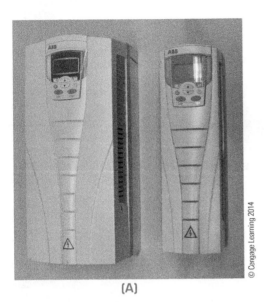

© Cengage Learning 2014

(A)   **(B)**

FIGURE 14–23 (A) Three-phase 480V Variable frequency drive. (B) Electronic circuit board inside a VFD.

INTRODUCTION TO PROGRAMMABLE CONTROLLERS

Modern motor control requires more exact timing and faster, more consistent operations than those provided by the older electromechanical relays and timers. Because modern manufacturing requirements also rely on system flexibility, a fast change system was needed that did not require extensive redesign, building, hand wiring, and testing to create a new motor control scheme.

Along with this change in needs, the microprocessor was improved to make it more reliable in industrial environments and more trouble-free in operation. The combination of these events produced the programmable logic controller (PLC).

Many versions of the PLC are made by various manufacturers and have various capabilities. However, all programmable controllers have at least three basic components. The first is a processor, which is a microprocessor that contains the instructions and makes the decisions. The second component is an input/output section, which receives input information from the process it is controlling and connects it to the controller in the proper format. It also takes the output from the controller and interfaces to the real world of control devices. The third main component is the programmer. This is the operator's access to the controller. The programs are written in ladder-logic style, which is familiar to most electricians. This familiarity was the factor that allowed the PLC to be so widely accepted as a control system. Controls were easily

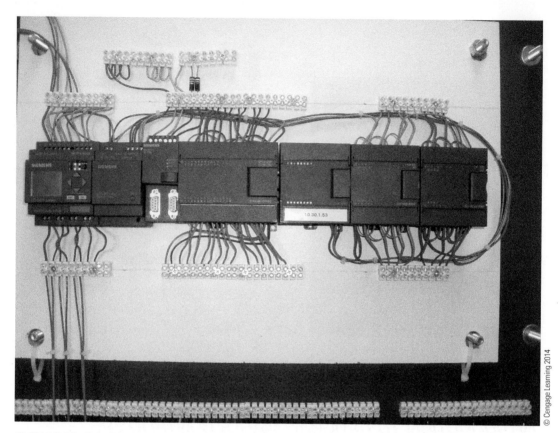

FIGURE 14-24 PC with one I/O card removed.

converted and adapted by working electricians, without them having to learn extensive new microprocessor programming language.

PLCs are used where the control system is likely to be changed frequently and usually where there are multiple functions for the controller to monitor, compare, count, time, or operate. These conditions make the use of a PLC economical and practical.

Large control systems may have various input/output cards that must be coordinated according to the "field wiring," such as AC and DC voltage at various levels, transistor–transistor logic (TTL), input/output (I/O), analog I/O, thermocouple, or binary-coded decimal (BCD). Figure 14–24 shows a PLC.

The intent of this book is not to teach programming of the PLC, but to familiarize the electrician with the possibility of motor control using the PLC. *Technician's Guide to Programmable Controllers,* 5th edition, by Richard A. Cox (Delmar Cengage Learning, 2006) is an excellent reference to gain further generic information.

© Cengage Learning 2014

SUMMARY

Many methods are used to start, stop, jog, and reverse three-phase motors. The basic operations generally use a magnetic controller to supply power to the motor. The reversing controller is used in the plugging operation to momentarily reverse the power and therefore bring the motor to a quick stop. Dynamic braking can also be used to stop an AC motor by applying DC to the motor field. Various methods of reducing the starting current to the motor employ the application of reduced voltage at the motor terminals during the starting period. Resistance or reactance can be inserted in series with the motor to reduce the voltage. An autotransformer may be used to reduce the applied voltage. Wound-rotor motors use secondary resistors to keep the starting current to a minimum. Wye–delta starters can be used to reduce the starting current to the motor by changing the configuration of the connections. Solid-state starters are now being installed to reduce the inrush current and to control the speed and the stopping characteristics.

Variable frequency drives use electronics to control the frequency to the motor and therefore control the speed of the motor. Many motor-control schemes can be developed with the use of a PLC. This electronic equivalent of a relay system is used to provide exact timing and complex, but changeable, control through the use of a microprocessor-based system.

ACHIEVEMENT REVIEW

1. What is meant by the phrase *jogging a motor?*

   _____

   _____

   _____

   _____

2. What is meant by the term *plugging?*

   _____

   _____

   _____

   _____

3. How is dynamic braking applied to an AC induction motor?

_____

_____

_____

_____

4. How is dynamic braking applied to a synchronous motor?

_____

_____

_____

_____

5. Draw a schematic diagram of the connections for the control circuit of an ATL magnetic motor switch with jogging capability. Include the main relay coil; the pushbutton station with start, jog, and stop pushbuttons; and the sealing contactor.

6. What identifying information should appear on a motor controller so that the controller complies with the requirements of the *National Electrical Code*®?

_____

_____

_____

_____

7. Draw a schematic diagram of an automatic controller used for a wound-rotor induction motor.

8. A three-phase AC induction motor has the following ratings: 15 hp, 230 volts, 42 amperes per terminal, 40° Celsius, and code classification F. In the spaces in the following table, insert the correct values for fuse protection and running overcurrent protection for this motor when used with each of the types of controllers listed.

	NONTIME DELAY	TIME DELAY	RUNNING PROTECTION
SYNCHRONOUS MOTOR			
WOUND-ROTOR MOTOR			
THREE-PHASE DESIGN B			

9. What is the purpose of an automatic autotransformer starting compensator?

_____

_____

_____

10. What is the purpose of an automatic controller used with wound-rotor induction motors? _____

_____

_____

_____

_____

11. A wye–delta controller starts the motor at _____
   a. 173% of the line voltage.
   b. 58% of the line voltage.
   c. full line voltage.
   d. 25% of the line voltage.

12. A three-phase, wye–connected motor _____

    a. should never be started by a wye–delta controller.

    b. should always be started by a wye–delta controller.

    c. can be started by a wye–delta controller if proper timers are used.

    d. can be started by a wye–delta controller if proper pushbuttons are used.

13. How are SCRs connected to pass and control AC? _____

_____

_____

_____

_____

14. If a solid-state controller has contactors in the power circuit, in what position are the contacts of start and run in the off position? Why? _____

_____

_____

_____

15. Why are adjustable speed drives used? _____

_____

_____

_____

16. List the types of AC adjustable speed drives.

_____

_____

_____

_____

_____

17. How is the speed of the wound-rotor motor adjusted?

_____

_____

_____

_____

_____

18. How is the eddy-current drive coupled to the load?

_____

_____

_____

_____

_____

19. How is the AC frequency varied in the mechanical method drive?

_____

_____

20. What is the formula for calculating AC motor speed? _____

_____

_____

21. What basic devices are provided in the adjustable frequency drive?

_____

_____

_____

22. With an apparent high degree of skill required to maintain an adjustable frequency drive control, how would a plant electrician quickly repair a defective unit?

_____

_____

_____

_____

_____

_____

_____

UNIT 15

SOLID-STATE STARTERS AND CONTROLS

OBJECTIVES

After studying this unit, the student should be able to

- use the correct terminology when selecting the proper motor-controller parameters.

- identify the operating characteristics of solid-state starters.

- determine the advantages and disadvantages of solid-state starters.

- determine the correct application of solid-state pilot devices.

- note the differences between National Electrical Manufacturers Association (NEMA) controllers and International Electrotechnical Commission (IEC) controllers.

In addition to the variable frequency speed controls covered in Unit 15, many motors are being installed with electronic controllers. The starters are more reliable than ever and have many features that are not available on conventional electromechanical starters. These starters can be used for a variety of applications, including pumps, blowers, compressors, chillers, and any other motor use. They can be used on induction, synchronous, multispeed, reversing, and even wound-rotor motors. DC motor drives are described in Unit 6.

DIGITAL CONTROLLERS

Many of the electronic solid-state starters now available have advanced features that allow the electrician to choose just the right parameters to fit the intended control need. With the spread of worldwide control companies, most starters are designed to operate over a range of supply voltages and frequencies. Many starters are programmable so that the field electrician can make adjustments to specially start, stop, and monitor the motor operations. This starter can be programmed using a remote digital display to set the controller for the desired motor full-load amperes (FLA). It allows the user to set the controller to the proper motor service factor as well as motor acceleration and deceleration. It also may protect the motor from phase imbalance or improper phase sequence. The electronic controller can monitor current to the motor and protect the motor from overload, selected as class 10, 20, or 30 overload. It may open the line power if one phase is lost on a three-phase line (a condition known as *single phasing*).

Features available on many digital solid-state starters include over- or under-voltage protection, or voltage-unbalance protection. Ground-fault protection may also be chosen as an option with many of the controllers. Many of the manufacturers also offer display readout of fault conditions with the sequence of faults and conditions. This helps in the troubleshooting procedures when trying to ascertain conditions that occurred when the motor stopped functioning normally.

The programmable parameters must be set when installing a new electronic drive. A common parameter is the motor FLA, which is set based on the motor nameplate rating. This setting is designed to program the controller for a specific motor and adjust the operating characteristics to meet that particular motor's operation. A motor overload parameter is set based on the motor service factor. If the motor is able to run with a percentage overload of the design horsepower, the controller needs the information to detect when a dangerous motor overload occurs. Service factors may range from 1.0 to 1.4. Service factor (SF) 1 (or 100%) means no overload is tolerated by the motor design. SF 1.4 means that the motor can withstand 140% overload without damage to the motor windings. The overload current sensors must have the appropriate information to track motor operations.

Starting current parameters are established to determine how much current is to be delivered to the motor during start-up. It should be set to allow the motor to start turning as soon as the start command is given. In some starters, this parameter is listed as a percentage of the FLA of the motor. If the motor does not begin turning when the start command is received, then increase the value in small increments. If the motor accelerates too quickly and the starting torque is too hard on the driven load, then try reducing the starting current parameter. The maximum starting current parameter must be set. This setting is designed to limit the current to a maximum value based on the electrical supply requirements. The motor may not require the maximum value, but the limit will keep too much current from being drawn from the electric supply. This control should be checked with the next control to make sure the motor achieves full speed during the time limit set by the ramp-up time parameter.

Motor ramp-up time is the setting that determines how much time is to elapse before the motor is at full speed with full voltage and current applied. It controls the time in seconds between initial motor current setting and the time when maximum motor current is applied to the starting motor. Associated with the ramp-up time is the motor stall-time parameter. The stall time determines the seconds allowed from the end of the ramp-up time until the motor reaches full operating speed before the controller disconnects the motor from the line.

Other parameters to select are the deceleration modes. Some controllers have multiple modes for deceleration. There may be an initial level of voltage applied to the motor when the stop command is issued. This allows the voltage to the motor to slowly decrease and provide gradual stopping. If you need the motor to come to a quick stop, set this selection to 0 or no ramp-down voltage. As in the increasing speed mode, or acceleration mode, the next parameter is programmed with consideration of the deceleration time setting made in the next step. The deceleration voltage level 2 is set to control the voltage after the ramp-down time has expired, but the motor is still not at rest. This step 2 sets the end of the ramp voltage that the motor will receive as it is decelerating. If the motor is still turning at the end of the ramp-down time, decrease the step 2 voltage setting. The time setting for the deceleration is the next parameter. Set this to the desired time that is required for the motor to decelerate. This is not a braking control.

Line current imbalance is another commonly set point. If the phase currents to the motor are different by anywhere from 5% to 40% and the difference occurs for more than 10 seconds, the motor will be disconnected from the line to prevent it from being damaged by the single-phasing condition.

BASIC OPERATIONS

Electronic controllers are packaged in a ready-to-use style (see Figure 15–1). They are sized in various categories to accommodate standard motor sizing. Some electronic controllers use the pulse-width–modulation (PWM) method of controlling output voltage and motor current; some use a variation of the PWM drive; and some of the newest controllers use a flux vector method of

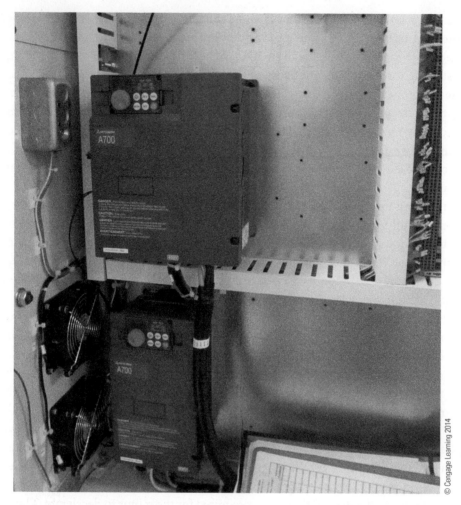

FIGURE 15–1 Solid-state motor controllers mounted in cabinets.

controlling the output to the motor. The difference between the two systems is in the feedback of motor operations to the controller and the sophistication of the microprocessor controller.

The PWM electronic drives with variable voltage control use a sequence of control that is simplified in Figure 15–2. In this block diagram, the incoming power is rectified and filtered in the initial stage of the controller with a three-phase rectifier circuit, as shown in Figure 15–3. A microprocessor controls the firing of the output controller, based on the parameters selected and the information from the controller sensors. The output is controlled by a configuration of transistors of a type called insulated gate bipolar transistor

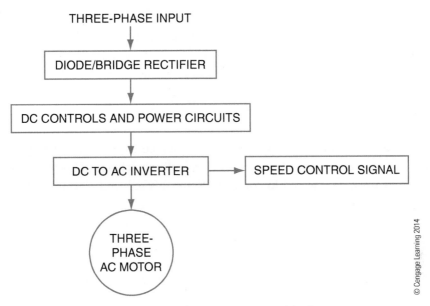

FIGURE 15–2 Block diagram of a typical AC variable-frequency electronic drive.

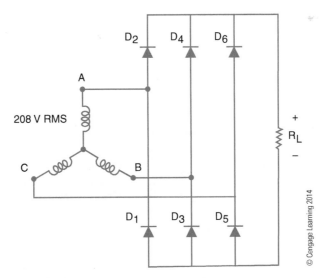

FIGURE 15–3 A wye source supplies power to a three phase, full-wave rectifier.

(IGBT). These electronic devices control the frequency and the amount of voltage that is delivered to the motor. By turning on the DC voltage for varying lengths of time, the pulses of DC are controlled and create an approximation of a sinewave output (see Figure 15–4). By adjusting the pulse width, or the time of pulses, and the modulation of the voltage available, PWM controls the frequency and the level of the output voltage supplied to the motor. A microprocessor responds to the input parameter selection and drives the motor to the design limits set by the operator.

Adjusting the frequency to an AC motor directly affects its speed. If the frequency applied is lower than the nameplate design frequency, the inductive reactance of the motor is lower, and the input voltage must be lowered to compensate for this reduced reactance of the winding. Line and output reactors may be included to reduce the disturbances to the power line. Because this is a high-speed switching device, it is considered a nonlinear load to the power source, and electrical harmonics are generated. The reactors help reduce the effects of the harmonics on the supply system. Note that there are several input points for digital inputs, such as contact closures, and the analog input points are typically for a speed-setting potentiometer. The output points are used to relay information to the user in the form of contact closure for indicating

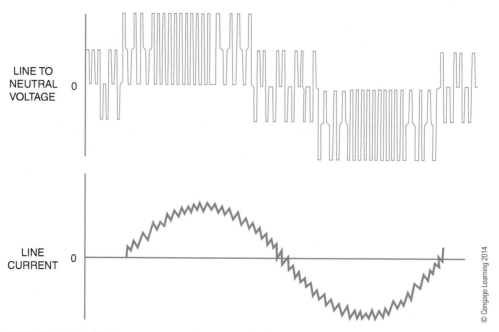

FIGURE 15–4 PWM output waveforms. Each phase has the same waveform.

© Cengage Learning 2014

lights or other needed alarms. The analog output can be used to drive other devices or as a monitor point for process control.

The flux-vector drive is an advancement in the PWM control for motors. If the motor is to run at set frequencies for extended periods and there are not a lot of load fluctuations, the speed stays nearly constant. If, however, there is a need for precise speed control and the load fluctuates considerably, then the flux-vector drive is needed. The advantage of the flux vector is derived from the feedback information to the controller from the output to the motor. It can produce maximum torque at all speeds, and the response to changing load is nearly instantaneous for speed changes. The term *vector* is used to describe the operating characteristics that react like a vector in mathematics. The vector describes direction and magnitude. The flux (magnetic fields) of the motor can be controlled like a vector to change magnitude and relative position of the magnetic fields in the motor. If the rotor flux and the stator flux can maintain an optimum 90° relationship to each other, then maximum torque is produced. This is the goal of adjusting the magnetic flux of the stator to create a 90° separation from the rotor flux.

A squirrel-cage motor produces rotor flux because it slips behind the stator magnetic field. The slip is measured in percent slip. Normally, the slip is 3% to 5% for a normal motor at full load. The slip can also be measured as an angular velocity in hertz. As the motor is loaded, the slip increases and the angular velocity increases, or the hertz slip increases. By monitoring the relative amount of flux in the stator (the magnitude) and the relative position of the flux (the direction of the vector), the controller uses a series of calculations to determine the exact speed and torque of the motor. The input and output sections are similar to the PWM drive. The controller is the difference. By sensing the magnitude of the current to the motor, a current feedback is created. By sensing speed in relation to the current, a speed feedback is created. The speed is monitored by a tachometer, an encoder, or a resolver. This signal is returned to the microprocessor controller, and then the PWM output is controlled to respond. The output is thus monitored to maintain set speed at a desired torque, and the field intensity and orientation are controlled for accurate speed control (see Figure 15–5).

Another form of the vector drive is called the *sensorless vector drive*. In this method, the manufacturer does not provide a direct connection to the motor through a speed sensor, but instead measures the different components of the current delivery to the motor. The concept of the flux-vector control is the same as in the sensor type of control, that is, to monitor the current that produces flux and the current that produces torque and calculate the angle between them. This combination of the two force vectors, magnetic flux vector and torque vector, creates the desired optimum combination of the two forces. Current sensors designed to measure the different vector currents are part of the control system. Figure 15–6 shows the basic control

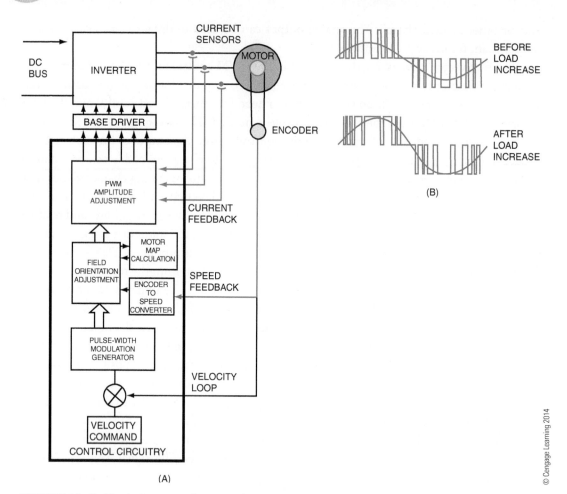

FIGURE 15–5 Block diagram of vector-drive control.

scheme. With the measurements, the microprocessor is used to calculate the slip and therefore can compensate for speed and torque changes by adjusting the output IGBTs.

CORRECT APPLICATION

When selecting a starter for a particular motor drive situation, you should consider several things. Is there a need for solid-state starting if the driven machinery of the electrical supply lines requires a form of reduced-voltage/reduced-current starting? Consider whether the driven machinery can start with reduced torque, compared to full voltage ATL starting. In some cases, the motor can be brought up to speed with an electronic drive, and then if it is to run at a constant speed, the electronics can be bypassed with an electromechanical starter. The advantages

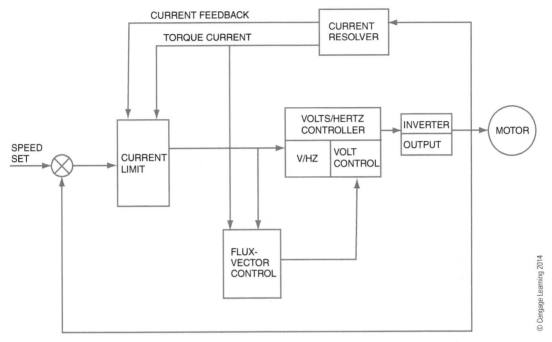

FIGURE 15–6 Basic block diagram of flux-vector controller.

to this method are that the electronics are not used on a continuous basis, and the heat gener-ated by the electronic switching does not have to be dissipated by the heat sinks. The efficiency of the drive increases because little power is lost in the electronic rectification and inverters back to the AC for the motor.

If the motor is to have multiple set speeds, the correct number of frequency set points must be determined and matched with the controller's capability. This is a feature that allows the user to set specific frequencies that the controller will hold for extended periods. If these are standard motor speeds, a consideration would be to use a simple control and a multispeed motor. Depending on the need for accuracy of the motor speed and the amount of control desired, various methods of control may be considered. A simple PWM drive is fairly accurate but does not use much feedback from the motor, whereas a vector drive has tighter control and more feedback from the motor operations.

SOLID-STATE SENSORS

Along with the evolution of the motor starter in the use of electronics for the control of power is the evolution of the sensor industry to feed signals to the controllers. The controllers may be programmable controllers (PC), which in turn supply a signal to an electromechanical starter or to an electronic controller. Whatever finally controls the motor speed and direction needs sensory input from the real world. These inputs can be light, heat, pressure, proximity, sound, or any other sense that starts some predetermined sequence of operation for the motor. Sensors now can be electronic rather than electromechanical in many cases. The limit switch, which has a feeler or roller as in Figure 15–7, can be replaced by an inductive or capacitive sensor, which does not require physical contact to sense an object's presence.

Proximity Sensors

Proximity sensors (controls) are available in two types: inductive sensors and capacitive sensors. A typical inductive sensor is an electronic circuit type of sensor that produces a small magnetic field at a radio frequency (RF). As the magnetic waves are produced, a tuned circuit is set to react to an altered oscillation created by a detected metal object. The tuning of these circuits is adjustable to account for the distance of detection and the amount of metal

© Cengage Learning 2014

FIGURE 15–7 Styles of contact style limit switches.

that is required before sensing is accomplished. As a metal object approaches the inductive proximity sensor, the magnetic coil in the sensor actually induces eddy currents into the detected target. This causes a change in the RF oscillations, and the amplitude on the RF is reduced or dampened. This dampened signal is detected by the driving electronics and switches the proximity sensor output to an "on," or dampened, state. As the object moves away from the sensor, the oscillator output returns to the undampened state, which allows the output to turn off.

Inductive proximity sensors work well in dirty or greasy environments where photo sensors do not work well. They have very fast response and can detect lightweight objects that cannot be detected by mechanical limit switches. Inductive sensors also can be used to detect metal objects behind nonmetallic materials such as glass or plastic. The two drawbacks to an inductive sensor are that the target to be detected must be metal, and the distance of detection is limited to approximately 4 in. The output of these sensors can be a relay contact or other type of output desired for the associated control circuitry (see Figure 15–8).

Capacitive sensors work because of the effects of different materials on the dielectric constant of a capacitive device. Various nonmetal materials have different dielectric constants. As a material is sensed by a capacitive sensor, the dielectric constant of the circuit changes and starts an oscillation that is detected by the receiver. This altered condition initiates the electronics to cause an output relay to change state (open to closed, or vice versa).

The capacitive proximity sensor has an advantage over the inductive sensor in that it can detect both nonmetallic and metallic objects. It has a larger sensing range and can be used for

FIGURE 15–8 Proximity sensors.

Courtesy of Rockwell Automation, Inc.

detecting liquid targets through nonmetallic barriers. It can be used for looking through glass or plastic tanks to detect liquid or other materials, such as plastic pellets in a feeder hopper.

The capacitive sensor has several limitations. It is more affected by moisture and humidity, and if embedded in the equipment, the surrounding material must be considered in the calibration. Also, because it has greater sensing range, more space must be maintained between sensors, and the target area must be well defined (see Figure 15–9).

Many types of sensors as well as many types of pilot devices are available. All are intended to provide input information to control electrical systems. Up to this point, we have only discussed simple on-off control schemes where the pilot device is the decision point to energize or de-energize a motor starter. If more complex controls are needed with many different inputs controlling many different output functions, another decision maker may be needed, such as a programmable controller.

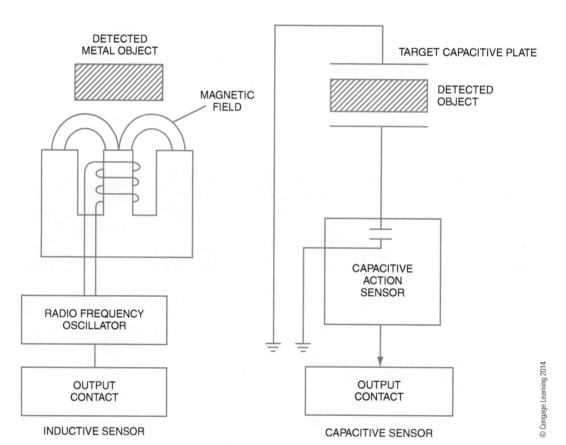

FIGURE 15–9 Simplified diagrams of inductive and capacitive proximity sensors.

© Cengage Learning 2014

Photoelectric Controls

Photoelectric sensors have a wide variety of methods to detect an object. Other than mechanical actuation, other types of sensors are also used. Photoelectric controls are used where physical contact is not needed or desired. Photoelectric control covers a broad spectrum of controls. The light used is not always visible. The light may be anywhere in the light spectrum from ultraviolet at one end to infrared at the other (see Figure 15–10).

The methods of gathering or sensing the light also may vary. Photoelectric controls use a variety of devices to detect the presence or absence of light.

Photodiode sensors are light-activated sensors with a semiconductor diode that uses a clear covering for the diode case. Because the diode is operated in reverse bias, it does not allow current to flow in the circuit when no light is striking the transparent case. However, as light energy increases on the diode, current is allowed to flow in the circuit. This, in turn, controls a transistor to provide the switching needed in the circuit.

Photoresistive cells (or photoconductive cells) also are used. Often the concept used in photoelectric control is employed to monitor surrounding daylight conditions and control outdoor lighting. When light strikes the photoresistive cell, the resistance of the device drops and allows a larger current to flow. The resultant current flow often picks up a relay (electronic relay) to provide the desired switching result. A cadmium sulfide cell (cad cell) is used because the resistance changes from approximately 50 ohms in direct sunlight to many thousands of ohms when dark.

When used in conjunction with more electronic circuitry, the switching point of the relay can be changed to reflect a desired light level control. Older types of photocells often used a metal hood to cover the cad cell. This allowed control of the amount of light on the cell and allowed adjustment of the activation point.

Caution: When installing photocells to control lighting based on available (ambient) light, make sure the cell faces away from the light source you are controlling. Otherwise, as dusk arrives (and you want to turn on the outdoor lighting), the cad cell looking at the light source turns off the lights it just turned on. The result is blinking lights that go on and off all night.

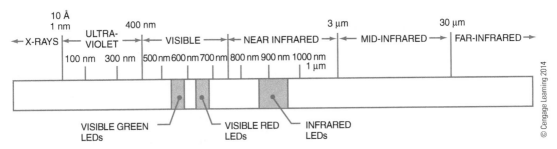

FIGURE 15–10 Light spectrum of photoelectric sensors.

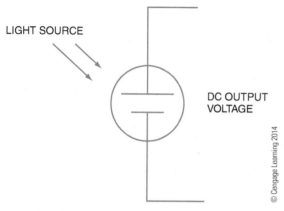

LIGHT SOURCE

DC OUTPUT
VOLTAGE

© Cengage Learning 2014

FIGURE 15–11 Photovoltaic cell produces
DC output voltage when struck by light.

Photovoltaic cells are sometimes also used. As the name implies, the cell actually creates or generates a voltage as the light strikes the cell. Current silicon photovoltaic cells (solar cells) produce approximately 0.5-volt DC at 150 mA for each square inch of cell area. The symbol in Figure 15–11 indicates that the cell actually produces a DC voltage and is useful where an external voltage source may not be convenient.

Sensor style and technique are different based on the type of sensor and application. Photosensors that use their own internal light source rather than ambient light use one of several styles of reflection techniques to "see" the light source. The items to be seen, the physical space, and the surrounding light help determine which reflective technique should be used.

Retroreflective scan sensors use a light source and receiver mounted in the same enclosure. The light is directed toward a reflective surface such as a common bicycle type of reflector (see Figure 15–12). The reflected light is detected by the receiver and in turn picks up a *relay*. Often these devices can be arranged to *pick light* or *pick dark*. Pick light means the relay is energized with a completed light beam. Pick dark means the relay is picked up when the light beam is obstructed and the light is not reflected.

RETROREFLECTIVE SENSING MODE

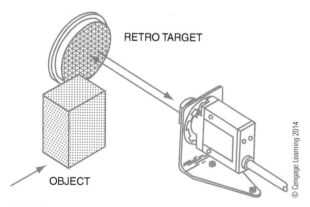

RETRO TARGET

OBJECT

© Cengage Learning 2014

FIGURE 15–12 Retroreflective sensor uses a light
source and a reflective target. Object to be sensed
is between sensor and reflector.

Retroreflective sensors are a good style to use when you have room for a reflector and the object to be sensed is not reflective itself. In addition, the alignment is not critical, so some vibration of the scanner or reflector is tolerated. The typical distance limit to the target reflector is approximately 40 ft. One advantage of the retroreflective type of sensor is that all wiring is to one device with a nonelectrical reflector placed where convenient. Retroreflective scan also could be used to count objects that have reflective tape applied. As the objects move past the sensor, the reflected light causes a contact closure in the retroreflective sensor.

Diffuse scan sensors use the principle of a light source directed at the surface of objects that are not reflective. Most of the light source is absorbed by the object, and the rest is diffused in all directions. A receiver is placed at about the same distance from the surface as the light source. The small amount of light that is returned to the receiver indicates presence of the object. Most diffuse scan sensors use a lens to collinate, or make the light rays parallel, so that the receiver can receive more returned light. These sensors are used in close proximity to the object to be detected (see Figure 15–13).

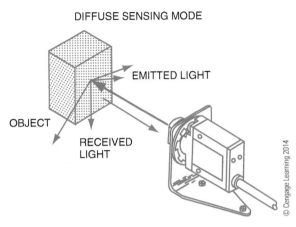

FIGURE 15–13 Diffuse mode sensor with source and receiver in the same housing.

A convergent beam is used to detect objects that are near other reflective surfaces, or for scanning small objects (see Figure 15–14). *Convergent beam* sensors use the light source and sensing device in a single head. The light beam is focused or converged at the object to be sensed. This is a fixed distance determined by the sensor and is adjustable by field focus controls. If the object is not at the focal point of the beam, no light is reflected back to the receiver. If an object is at the focal point (not in front of or behind the focal point), a small amount of light is reflected to the receiver. Only one head is used and the surrounding surfaces are

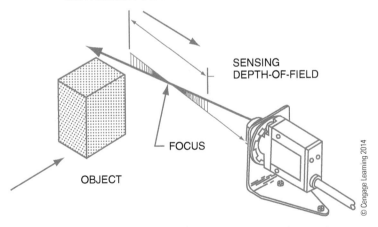

FIGURE 15–14 Convergent beam sensor. Light is focused at a specific point where the object is sensed.

SPECULAR SENSING MODE SENSES THE DIFFERENCE
BETWEEN SHINY AND DULL SURFACES

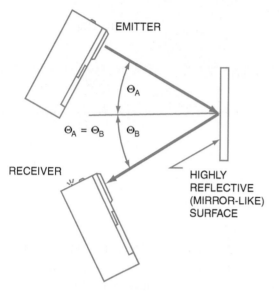

FIGURE 15–15 Specular scanner uses an emitter and receiver to sense objects with shiny surfaces.

ignored. Convergent beam sensors are ideal where the objects are near other surfaces or where the objects have low reflectivity.

Specular scan sensors use a light source and a receiver mounted at angles to each other, as in Figure 15–15. The principle is to use a reflected light beam off a shiny object. The focal point is set to reflect light from the light source to the receiver. If an object interferes with that reflection, the receiver "sees" the object. Another option is for the object to complete the reflected beam. By adjusting the angle of reflection, a shiny object can cause the beam to be reflected back to the receiver to "see" the object. Mounting angles are very precise, and a lot of vibration is not tolerated.

The least complicated sensor is called *direct scan*. This method uses a transmitter (light source) and a receiver, as shown in Figure 15–16. If an object passes between the transmitter and receiver, the beam is broken and the object is detected. Alignment is critical in the sensors,

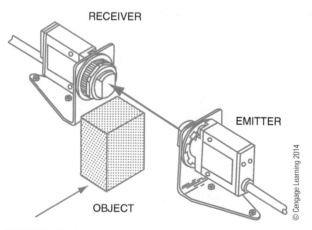

FIGURE 15–16 Direct scan or opposed sensing uses an emitter and receiver to detect an object between them.

and wiring is required to both the transmitter and receiver. Direct-scan sensors are harder to keep in alignment, but they are used extensively.

Many light sources use visible light, but the options include ultra violet light and infrared light. When sources other than visible light are used, aids are required to help set up, aim, and troubleshoot the devices. To determine whether an infrared light source is emitting, a small portable receiver is used to detect light output. If the light receiver is suspected to be defective, a portable light source can be used to transmit a known beam to the receiver. After that, troubleshooting regarding focus, improper application, or replacement of the defective device can proceed.

Fiber optics are also used to sense objects. Light is transmitted through a fiber-optic cable to the point where an object is to be detected. This allows the sensor to be located out of the way, and only the small cables have to be installed at the sensing location. Bifurcated fiber-optic cable uses the same fiber bundle to send and receive light (see Figure 15–17). A "thru" scan has a transmitter and receiver cable as separate cables.

Be sure to check the manufacturer's specifications when determining the proper photoelectric device to use. Specific application is important. Ambient light, temperature, vibration, moisture, and surrounding dust are all important considerations.

Strain gauges sense pressure and therefore can measure weight. An electronic strain gauge is typically made up of resistive components built into a wheatstone bridge configuration. The four parts of the bridge are shown in Figure 15–18. By bonding the semiconductors to a pressure-sensitive diaphragm, a change in expansion or contraction causes the resistance of the semiconductor to change. With a supply voltage applied, the output voltage is proportional to the change of the bridge resistance. With no pressure applied, the output is zero volts. As the pressure on the electronic elements changes, two of them increase resistance, and the other two decrease resistance. The output voltage proportionally changes to measure compression. The analog output voltage can then be sent to a monitor or controller to initiate some action. These strain gauges often have temperature stabilizers and offset adjustments to allow for calibration.

BIFURCATED FIBER OPTIC

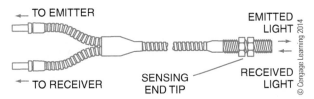

FIGURE 15–17   Bifurcated cable can be used in opposed mode. It uses fiber optics to bring light from the sensor head to the application.

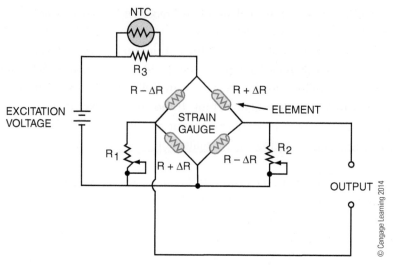

FIGURE 15-18 A semiconductor strain gauge configuration.

Temperature detection for industrial control is usually done through thermistors or resistive temperature detectors (RTDs). Thermistors are resistors that change resistance due to thermal effects, known as *thermal resistors* (shortened to *thermistors*). If the thermistor has a negative temperature coefficient, it means that as the temperature of the semiconductor rises, the resistance has a negative effect, or decreases (noted as NTC in Figure 15–19). If the resistance were to rise with an increase in temperature, the component would have a positive temperature coefficient. The resistors can be used directly in a circuit, which changes current flow as the resistance changes, or can be placed in a bridge much like a strain gauge to control an output voltage.

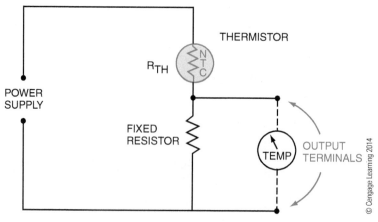

FIGURE 15-19 Basic thermistor temperature detector circuit.

An RTD typically uses metals that exhibit a positive temperature coefficient. Nickel and platinum are commonly used as the metals in the RTD sensing point. Nickel is used to sense temperature because it has a large change in resistance with small changes in temperature. Platinum is used when the temperature variations are large and therefore the ratio of the temperature to change of resistance is large. The RTD assembly is usually encased in a sensor style, and only the connection leads are available for connection.

IEC STANDARDS

To compete in the world market, U.S. manufacturers must produce goods that meet the technical and consumer requirements of other countries. The emphasis in Europe, for example, has been to produce electrical products that are more specific to need. This reduces the usage range and the amount of materials needed for the starter. Although customers of such products do not pay for more than they need, the products are not as rugged under adverse conditions.

The International Electrotechnical Commission (IEC), headquartered in Geneva, Switzerland, is a counterpart to the National Electrical Manufacturers Association (NEMA). IEC standards can be used to select the most suitable controller. The IEC philosophy is that utilization is a critical part of the selection process. The user must choose the utilization category, and then select a product in that category that fits the need. In addition, the user must check to determine if the manufacturer's contact life rating is adequate for the purpose. Typical utilization categories are presented in Figure 15–20. Contact life is also taken into account. To make the proper selection, use the proper application selection graphs, as shown in Figure 15–21.

The horizontal axis (Ie) in Figure 15–21 represents the operational current that the power contacts must handle. The vertical axis represents the number of operating cycles that can be expected in the contactor life. Read along the Ie axis to the point of the normal full-load current,

Common Utilization Categories for AC Contactors	
Utilization category	**Typical duty**
AC1	Noninductive or slightly inductive loads
AC2	Starting of slip-ring motors
AC3	Starting of squirrel-cage motors and switching off only after the motor is up to speed
AC4	Starting of squirrel-cage motors with inching and plugging duty.
Note: In an AC3 application, the contactor will never interrupt more than the motor's full-load current. If the application requires interruption of current greater than motor FLC, it is an AC4 application.	

Courtesy of Rockwell Automation, Inc.

FIGURE 15–20 Utilization list for typical duties of IEC starters.

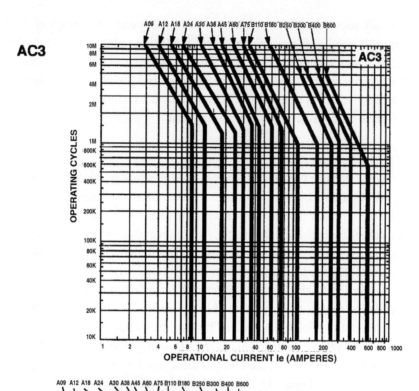

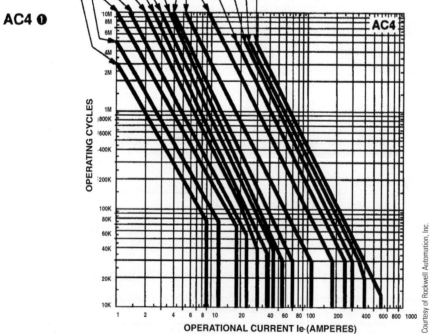

FIGURE 15-21 Use of charts in utilization categories helps in selection of the proper starter.

Note that the AC4 life load curves are based on the assumption that motors used have a locked rotor current equal to or less than 600% of motor full-load current.

and then up to the point a particular contactor crosses the current line. Reading horizontally over to the left supplies the projected number of operations. For example, the AC3 category is for starters that control squirrel-cage motors and that switch off only after the motor is up to speed. A 5 hp, 460-volt, three-phase motor draws 15.2 amperes according to NEC® *Table 430.250*. Using this data, read 15.2 amperes on the Ie axis. Read straight up to where the bold line crosses the A18 vertical line. Contactor A18 would provide approximately 1.5 million operations. If that does not seem sufficient, go up to the A24 bold line crossing the 15-ampere vertical line. An A24 may provide 3 million operations.

Certain considerations must be taken into account when selecting NEMA or IEC standards for installations. How much expertise do you need to make the proper selection? Do you want to dedicate the increased space usually needed for NEMA controllers? Do you have a good understanding of the application? If you are replacing an existing installation, what other components have to be changed? How much is cost a factor compared to interchangeability? Are overloads adjustable or changeable? These questions, along with the difference in price, must be considered when determining the proper selection.

SUMMARY

The advancement of electronics into the control of electrical high-power equipment has many facets. From the measurement of the real world and the collection of measurements, the information can be sent to control the power circuits of large motors through the use of high-power transistors. To correctly design, build, test, or troubleshoot a complex system, a solid knowledge of all the various components is necessary. To properly retrofit a failed system or diagnose problems in a malfunctioning system, a working knowledge of terminology and capabilities is necessary. Solid-state drives have many applications and in many cases can replace obsolete mechanical control systems. Solid-state pilot devices can also be used to resolve some of the toughest sensor problems.

ACHIEVEMENT REVIEW

A.   Answer the statements and questions in items 1 through 7.

1.   When one of the three phase conductors to a motor is disconnected, the condition that results is known as _____.

2.   Name two of the fault conditions that can often be detected by electronic drives.

_____

_____

3. Are motor ramp-up times always equal to motor ramp-down times? Explain.

_____

_____

_____

_____

4. A flux-vector drive has (better/worse) speed control than a standard PWM drive. (Circle the correct answer.)

5. Name two types of proximity sensors.

_____

_____

6. What does RTD stand for? _____

7. What are the two vector components that are measured to make a vector drive effective?

_____

_____

_____

B. Select the correct answer for the following statements, and place the corresponding letter in the space provided.

8. When programming a drive, the service factor must be known. SF of 1.0 means that the motor _____
   a. is rated for 1 hp.
   b. has a safety factor of 10 times the normal load.
   c. cannot be overloaded.
   d. needs to be serviced once per year.

9. PWM drives are actually electronic drives that output _____
   a. true sinewaves.
   b. an equivalent sinewave.
   c. six-step AC wave approximations.
   d. DC positive wave magnetics.

10. Strain gauges are used to measure _____
   a. temperature at a junction.
   b. relative tension between electrical signals.
   c. the pressure of compression or extension.
   d. the thickness of metal at the point of fatigue.

UNIT

THE WOUND-ROTOR INDUCTION MOTOR

16

BJECTIVES

After studying this unit, the student should be able to

- list the main components of a wound-rotor, poly-phase induction motor.

- describe how the synchronous speed is developed in this type of motor.

- describe how a speed controller connected to the brushes of the motor provides a variable speed range for the motor.

- state how the torque, speed regulation, and operating efficiency of the motor are affected by the speed controller.

- demonstrate how to reverse the direction of rotation of a wound-rotor induction motor.

Until the past several years, AC variable speed control was very difficult with a standard motor. Therefore, a different type of motor and control system was developed and used extensively for years. This motor is rarely installed as a new system now. Maintenance electricians must be familiar with this type of motor and control system.

Many industrial motor applications require three-phase motors with variable speed control. The squirrel-cage induction motor cannot be used without additional controls for variable speed work, because its speed is essentially constant. Another type of induction motor was developed for variable speed applications. This motor is called the *wound-rotor induction motor* or *slip-ring AC motor.*

WOUND-ROTOR AC MOTORS

Wound-rotor motor drives use a specially constructed AC motor to accomplish speed control and for their inherent ability to provide high starting torque with relatively low starting current. These motors are now installed in heavy manufacturing needs such as ball mills, shredders, and cement mills, to provide a rugged motor with the ability to bring high-inertia loads up to speed smoothly. The windings of the motor rotor are brought out of the motor through slip rings on the rotor shaft. Figure 16–1 shows an elementary diagram of a wound-rotor motor with an adjustable speed drive. These windings are connected to a controller that places variable resistors in series with the windings. The torque performance of the motor can be controlled using these variable resistors or liquid rheostats.

Wound-rotor motors are more common in the larger sizes, that is, 100 to 1000 hp and above.

Features of Wound-Rotor Motors

Wound-rotor motors have the following advantages, which make them suitable for a variety of applications:

- *Cost*—The initial cost is moderate for the high horsepower units.

- *Control*—Not all the power need be controlled, resulting in a moderate size and simple controller.

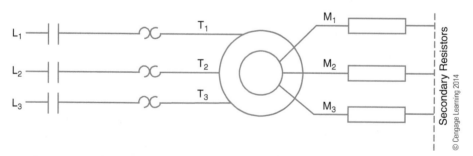

FIGURE 16–1 Elementary diagram of an adjustable speed drive, wound-rotor motor.

- *Construction*—The simple construction of the motor and control lends itself to maintenance without the need for a high level of training.
- *High-inertia loads*—The drive works well on high-inertia loads.

Disadvantages of Wound-Rotor Motors

Wound-rotor motors also have disadvantages:

- *Custom motor*—The motor has a rotor wound with wire and slip rings, and is not easily available, but is still manufactured.
- *Efficiency*—The drive does not maintain a high efficiency at low speeds.
- *Speed range*—The drive usually is limited to a speed range of two to one.

CONSTRUCTION DETAILS

A wound-rotor induction motor consists of a stator core with a three-phase winding, a wound rotor with slip rings, brushes and brush holders, and two end shields to house the bearings that support the rotor shaft.

Figures 16–2, 16–3, 16–4, and 16–5 show the basic parts of a wound-rotor induction motor.

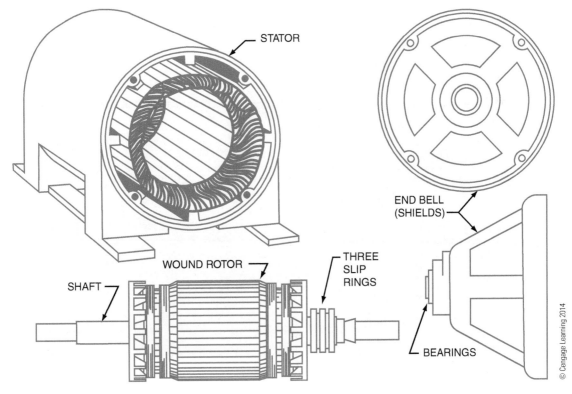

FIGURE 16–2 Parts of a wound-rotor motor.

© Cengage Learning 2014

FIGURE 16–3 Wound starter for a polyphase induction motor.

FIGURE 16–4 Rotor of a wound rotor motor.

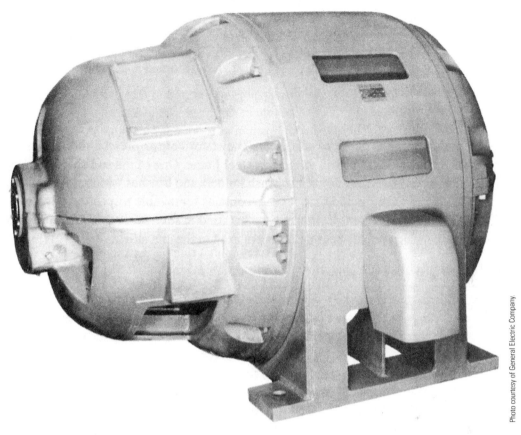

Photo courtesy of General Electric Company

FIGURE 16–5 Sleeve-bearing, wound-rotor, polyphase induction motor.

The Stator

A typical stator contains a three-phase winding held in place in the slots of a laminated steel core, shown in Figure 16–3. The winding consists of formed coils arranged and connected so that three single-phase windings are spaced 120 electrical degrees apart. The separate single-phase windings are connected either in wye or delta. Three line leads are brought out to a terminal box mounted on the frame of the motor. This is the same construction as the squirrel-cage motor stator.

The Rotor

The rotor consists of a cylindrical core composed of steel laminations. Slots cut into the cylindrical core hold the formed coils of wire for the rotor winding.

The rotor winding consists of three single-phase windings spaced 120 electrical degrees apart. The single-phase windings are connected either in wye or delta. (The rotor winding must have the same number of poles as the stator winding.) The three leads from the three-phase rotor winding terminate at three slip rings mounted on the rotor shaft. Leads from carbon brushes that ride on these slip rings are connected to an external speed controller to vary the rotor resistance for speed control.

The brushes are held securely to the slip rings of the wound rotor by adjustable springs mounted in the brush holders. The brush holders are fixed in one position. For this type of motor, it is not necessary to shift the brush position as is sometimes required in DC generator and motor work.

The Motor Frame

The motor frame is made of cast steel. The stator core is pressed directly into the frame. Two end shields are bolted to the cast steel frame. One of the end shields is larger than the other because it must house the brush holders and brushes that ride on the slip rings of the wound rotor. In addition, it often contains removable inspection covers.

The bearing arrangement is the same as that used in squirrel-cage induction motors. Either sleeve bearings or ball-bearing units are used in the end shields.

PRINCIPLE OF OPERATION

When three currents, 120 electrical degrees apart, pass through the three single-phase windings in the slots of the stator core, a rotating magnetic field is developed. This field travels around the stator. The speed of the rotating field depends on the number of stator poles and the frequency of the power source. This speed is called the *synchronous speed* and is determined by applying the formula used to find the synchronous speed of the rotating field of squirrel-cage induction motors.

$$\text{Synchronous speed in RPM} = \frac{120 \times \text{Frequency in hertz}}{\text{Number of poles}}$$

or

$$\text{RPM} = \frac{120 \times f}{p}$$

As the rotating field travels at synchronous speed, it cuts the three-phase winding of the rotor and induces voltages in this winding. The rotor winding is connected to the three slip rings mounted on the rotor shaft. The brushes riding on the slip rings connect to an external wye-connected group of resistors (speed controller), shown in Figure 16–6. The induced voltages in the rotor windings set up currents that follow a closed path from the rotor winding to the wye-connected speed controller. The rotor currents create a magnetic field in the rotor core based on

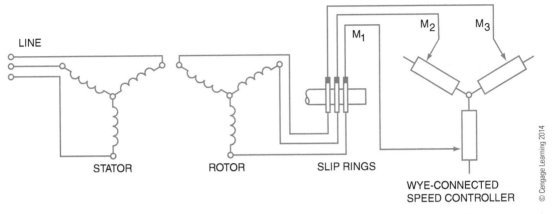

FIGURE 16–6 Connections for a wound-rotor induction motor and a speed controller.

transformer action. This rotor field reacts with the stator field to develop the torque that causes the rotor to turn. The speed controller is sometimes called the *secondary resistance control.*

Starting Theory of Wound-Rotor Induction Motors

To start the motor, all resistance of the wye-connected speed controller is inserted in the rotor circuit. The stator circuit is energized from the three-phase line. The voltage induced in the rotor develops currents in the rotor circuit. The rotor currents, however, are limited in value by the resistance of the speed controller. As a result, the stator current also is limited in value. In other words, to minimize the starting surge of current to a wound-rotor induction motor, insert the full resistance of the speed controller in the rotor circuit. The starting torque is affected by the resistance inserted in the rotor secondary. With resistance in the secondary, the power factor of the rotor is high or close to unity. This means that the rotor current is nearly in phase with the rotor-induced voltage. If the rotor current is in phase with the rotor-induced voltage, then the rotor magnetic poles are being produced at the same time as the stator poles. This creates a strong magnetic effect, which creates a strong starting torque. As the motor accelerates, steps of resistance in the wye-connected speed controller can be cut out of the rotor circuit until the motor accelerates to its rated speed.

Speed Control

The insertion of resistance in the rotor circuit not only limits the starting surge of current but also produces a high starting torque and provides a means of adjusting the speed. If the full resistance of the speed controller is inserted into the rotor circuit when the motor is running, the rotor current decreases and the motor slows down. As the rotor speed decreases, more voltage is induced in the rotor windings, and more rotor current is developed to create the necessary torque at the reduced speed.

If all resistance is removed from the rotor circuit, the current and motor speed increase. However, the rotor speed is always less than the synchronous speed of the field developed by the stator windings. Recall that this fact also is true of the squirrel-cage induction motor. The speed of a wound-rotor motor can be controlled manually or automatically with timing relays, contactors, and electronic speed control.

Through the use of solid-state controls, the wound-rotor motor can be started with the full secondary resistance in the circuit, and then the input power is also controlled to provide smooth acceleration and maximum torque. As the motor reaches full input voltage, the secondary resistors can be removed from the circuit, and the motor will operate with similar characteristics to a squirrel-cage motor. When resistance is reinserted in the secondary circuit, the motor speed slows, and the primary electronic drive can also be adjusted to provide gradual speed adjustment.

Torque Performance

As a load is applied to the motor, both the percent slip of the rotor and the torque developed in the rotor increase. As shown in the graph in Figure 16–7, the relationship between the torque and percent slip is practically a straight line.

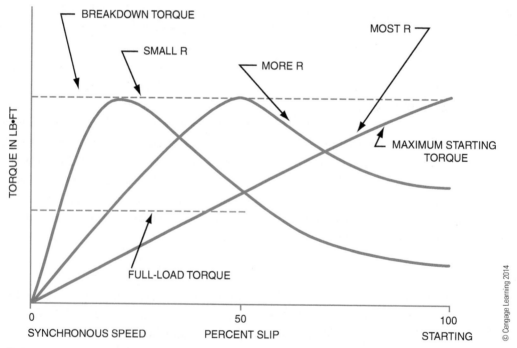

FIGURE 16–7 Performance curves of a wound-rotor motor.

Figure 16–7 illustrates that the torque performance of a wound-rotor induction motor is good whenever the full resistance of the speed controller is inserted in the rotor circuit. The large amount of resistance in the rotor circuit causes the rotor current to be almost in phase with the induced voltage of the rotor. As a result, the field set up by the rotor current is almost in phase with the stator field. If the two fields reach a maximum value at the same instant, there will be a strong magnetic reaction resulting in a high torque output.

However, if all speed controller resistance is removed from the rotor circuit and the motor is started, the torque performance is poor. The rotor circuit minus the speed controller resistance consists largely of inductive reactance. This means that the rotor current lags behind the induced voltage of the rotor, and thus the rotor current lags behind the stator current. As a result, the rotor field set up by the rotor current lags behind the stator field that is set up by the stator current. The resulting magnetic reaction of the two fields is relatively small, because they reach their maximum values at different points. In summary, then, the starting torque output of a wound-rotor induction motor is poor when all resistance is removed from the rotor circuit.

Speed Regulation

As discussed previously, inserting resistance at the speed controller improves the starting torque of a wound-rotor motor at low speeds. However, there is an opposite effect at normal speeds. In other words, the speed regulation of the motor is poorer when resistance is added in the rotor circuit at a higher speed. For this reason, the resistance of the speed controller is removed as the motor comes up to its rated speed.

Figure 16–8 shows the speed performance of a wound-rotor induction motor. Note that the speed characteristic curve resulting when all the resistance is cut out of the speed controller indicates relatively good speed regulation. The second speed characteristic curve, resulting when all the resistance is inserted in the speed controller, has a marked drop in speed as the load increases. This indicates poor speed regulation.

Power Factor

The power factor of a wound-rotor induction motor at no load is as low as 15% to 20% lag. However, as load is applied to the motor, the power factor improves and increases to 85% to 90% lag at rated load.

Figure 16–9 shows a graph of the power factor performance of a wound-rotor induction motor from a no-load condition to full load. The low lagging power factor at no load is because the magnetizing component of load current is such a large part of the total motor current. The magnetizing component of load current magnetizes the iron, causing interaction between the rotor and the stator by mutual inductance.

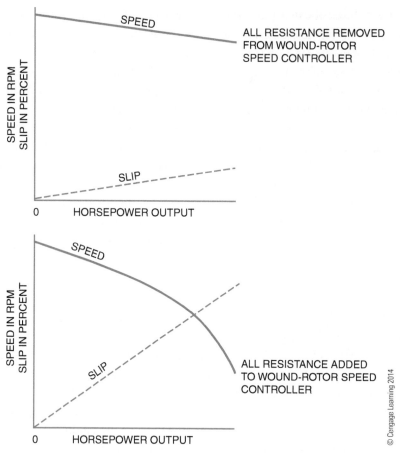

FIGURE 16-8 Speed performance curves of a wound-rotor motor.

As the mechanical load on the motor increases, the in-phase component of current increases to supply the increased power demands. The magnetizing component of the current remains the same, however. Because the total motor current is now more nearly in phase with the line voltage, there is an improvement in the power factor.

Operating Efficiency

Both a wound-rotor induction motor with all the resistance cut out of the speed controller and a squirrel-cage induction motor show nearly the same efficiency performance. However, when a motor must operate at slow speeds with all the resistance cut in the rotor circuit, the efficiency of the motor is poor because of the power loss in watts in the resistors of the speed controller.

Figure 16–10 illustrates the efficiency performance of a wound-rotor induction motor. The upper curve shows the highest operating efficiency results when the speed controller is in the fast position and no resistance is inserted in the rotor circuit. The lower curve shows a lower operating efficiency. This occurs when the speed controller is in the slow position and all controller resistance is inserted in the rotor circuit.

Reversing Rotation

The direction of rotation of a wound-rotor induction motor is reversed by interchanging the connections of any two of the three line leads, shown in Figure 16–11. This procedure is identical to the procedure used to reverse the direction of rotation of a squirrel-cage induction motor.

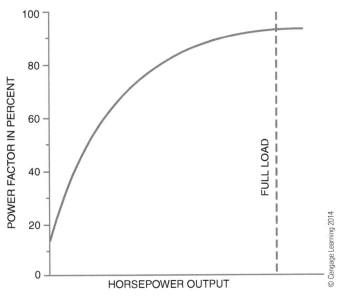

FIGURE 16–9 Power factor of a wound-rotor induction motor.

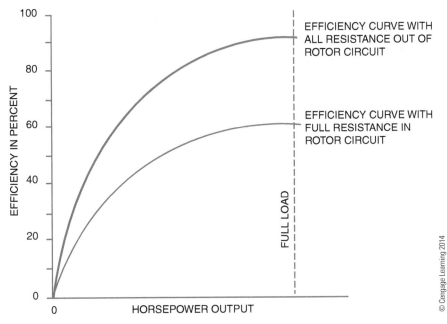

FIGURE 16–10 Efficiency curves for a wound-rotor induction motor.

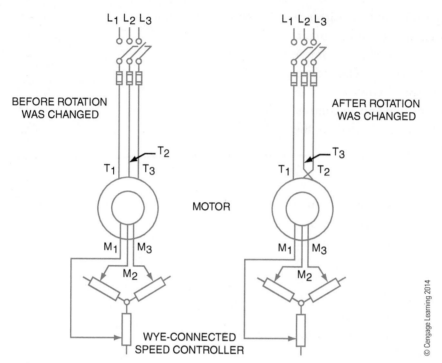

FIGURE 16–11 Changes necessary to reverse direction of rotation of a wound-rotor motor.

The electrician should never attempt to reverse the direction of rotation of a wound-rotor induction motor by interchanging any of the leads feeding from the slip rings to the speed controller. Changes in these connections will not reverse the direction of rotation of the motor.

SUMMARY

The wound-rotor motor is rarely installed as a new motor today, but many of these motors are still in use. Wound-rotor motors are typically larger horsepower motors assigned to special applications. The wound-rotor motor may be used for variable speed with the insertion of secondary resistors. The starting current and starting torque of the motor were the prime considerations when selecting the wound-rotor motor for installation. There are still many references to the wound-rotor motor used in the *National Electrical Code®*.

ACHIEVEMENT REVIEW

A. Provide complete answers to questions 1 through 9.

1. List the essential parts of a wound-rotor induction motor. _____

    _____

    _____

    _____

    _____

    _____

2. List two reasons why a wound-rotor induction motor is started with all resistance
    inserted in the controller. _____

    _____

    _____

    _____

    _____

3. A wound-rotor induction motor has six poles and is rated at 60 hertz. The full-load
    speed of this motor with all the resistance cut out of the speed controller is 1120 RPM.
    What is the synchronous speed of the field set up by the stator windings? _____

    _____

    _____

    _____

    _____

    _____

4. Determine the percent slip at the rated load for the motor in question 3.

_____

_____

_____

_____

5. Why is a wound-rotor induction motor used in place of a squirrel-cage induction motor for some industrial applications?

_____

_____

_____

_____

6. Why is the percent efficiency of a wound-rotor induction motor poor when operating at rated load with all resistance inserted in the speed controller?

_____

_____

_____

_____

7. What must be done to reverse the direction of rotation of a wound-rotor induction motor?

_____

_____

_____

_____

8. Why is the power factor of a wound-rotor induction motor poor at no load?

_____

_____

_____

_____

9. List the two factors that affect the synchronous speed of the rotating magnetic field set up by the current in the stator windings.

_____

_____

_____

_____

B. Select the correct answer for the statements in items 10 through 17, and place the corresponding letter in the space provided.

10. The speed of a wound-rotor motor is increased by _____
   a. inserting resistance in the primary circuit.
   b. inserting resistance in the secondary circuit.
   c. decreasing the resistance in the secondary circuit.
   d. decreasing the resistance in the primary circuit.

11. The starting current of a wound-rotor induction motor is limited by _____
   a. decreasing the resistance in the primary circuit.
   b. decreasing resistance in the secondary circuit.
   c. inserting resistance in the primary circuit.
   d. inserting resistance in the secondary circuit.

12. The direction of rotation of a wound-rotor motor is changed by interchanging any two of the three: _____
   a. L_1, L_2, or L_3.
   b. T_1, T_2, or T_3.
   c. M_1, M_2, or M_3.
   d. all of the above

13. Wound-rotor motors can be used with _____
    a. manual speed controllers.
    b. automatic speed controllers.
    c. pushbutton selection.
    d. all of the above

14. The full-load efficiency of a wound-rotor motor is best when _____
    a. all resistance is cut out of the secondary circuit.
    b. all resistance is cut in the secondary circuit.
    c. it is running slowly.
    d. it is running at medium speed.

15. The main advantage of the wound-rotor polyphase motor is that it _____
    a. has a low starting torque.
    b. has a wide speed range.
    c. will reverse rapidly.
    d. has a low speed range.

16. The wound-rotor motor is so named because _____
    a. the rotor is wound with wire.
    b. the stator is wound with wire.
    c. the controller is wound with wire.
    d. all of the above are true

17. The magnetizing component of load current _____
    a. is a small part of the total motor current at no load.
    b. magnetizes the iron, causing interaction between the rotor and the stator.
    c. is a large part of the total motor current at full load.
    d. is unrelated to the power factor.

UNIT

THE SYNCHRONOUS MOTOR

BJECTIVES

After studying this unit, the student should be able to

- list the basic parts in the construction of a synchronous motor.

- define and describe an amortisseur winding.

- describe the basic operation of a synchronous motor.

- describe how the power factor of a synchronous motor is affected by an under excited DC field, a normally excited DC field, and an overexcited DC field.

- list at least three industrial applications of the synchronous motor.

Courtesy of General Electric Company

FIGURE 17–1 Synchronous motor under construction.

The *synchronous motor*, shown in Figure 17–1, is a three-phase AC motor that operates at a constant speed from a no-load condition to full load. This type of motor has a revolving field that is separately excited from a DC source. In this respect, it is similar to a three-phase AC generator. If the DC field excitation is changed, the power factor of a synchronous motor can be varied over a wide range of lagging and leading values.

The synchronous motor is used in many industrial applications because of its fixed-speed characteristic over the range from no load to full load. This type of motor is also used to correct or improve the power factor of three-phase AC industrial circuits, thereby reducing operating costs.

CONSTRUCTION DETAILS

A three-phase synchronous motor basically consists of a stator core with a three-phase winding (similar to an induction motor), a revolving DC field with an auxiliary or amortisseur winding and slip rings, brushes and brush holders, and two end shields housing the bearings that support the rotor shaft. An *amortisseur winding* (Figure 17–2) consists of copper bars embedded in the cores of the poles. The copper bars of this special type of squirrel-cage winding are welded to end rings on each side of the rotor.

AMORTISSEUR
(SQUIRREL-CAGE
WINDING)

SLIP RINGS

SALIENT POLES

Photo courtesy of General Electric Company.

FIGURE 17-2 A synchronous motor with amortisseur winding.

Both the stator winding and the core of a synchronous motor are similar to those of the three-phase AC induction motor and the wound-rotor induction motor. The leads for the stator winding are marked T_1, T_2, and T_3 and terminate in an outlet box mounted on the side of the motor frame.

The rotor of the synchronous motor has salient field poles. The field coils are connected in series for alternate polarity. The number of rotor field poles must equal the number of stator field poles. The field circuit leads are brought out to two slip rings mounted on the rotor shaft for brush-type motors. Carbon brushes mounted in brush holders make contact with the two slip rings. The terminals of the field circuit are brought out from the brush holders to a second terminal box mounted on the frame of the motor. The leads for the field circuit are marked F_1 and F_2. A squirrel-cage, or amortisseur, winding is provided for starting because the synchronous motor is not self-starting without this feature. The rotor shown in Figure 17–2 has salient poles and an amortisseur winding.

Two end shields are provided on a synchronous motor. One of the end shields is larger than the second shield because it houses the DC brush-holder assembly and slip rings. Either sleeve bearings or ball-bearing units are used to support the rotor shaft. The bearings are also housed in the end shields of the motor.

PRINCIPLE OF OPERATION

When the rated three-phase voltage is applied to the stator windings, a rotating magnetic field is developed. This field travels at the synchronous speed. As stated in previous units, the synchronous speed of the magnetic field depends on the frequency of the three-phase voltage and the number of stator poles.

The magnetic field that is developed by the stator windings travels at synchronous speed and cuts across the squirrel-cage winding of the rotor. Both voltage and current are induced in the bars of the rotor winding. The resulting magnetic field of the amortisseur (squirrel-cage) winding reacts with the stator field to create a torque that causes the rotor to turn.

The rotation of the rotor increases in speed to a point slightly below the synchronous speed of the stator, about 92% to 97% of the motor rated speed. There is a small slip in the speed of the rotor behind the speed of the magnetic field set up by the stator. In other words, the motor is started as a squirrel-cage induction motor.

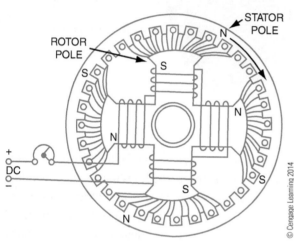

FIGURE 17–3 Diagram showing the principle of operation of a synchronous motor.

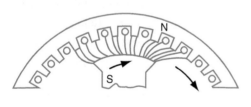

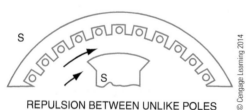

REPULSION BETWEEN UNLIKE POLES

FIGURE 17–4 Starting of synchronous motors.

The field circuit is now connected to a source of direct current, and fixed magnetic poles are set up in the rotor field cores. The magnetic poles of the rotor are attracted to the unlike magnetic poles set up by the stator magnetic field.

Figure 17–3 and Figure 17–4 show how the rotor field poles lock with unlike poles of the stator field. After the field poles are locked, the rotor speed becomes the same as the speed of the magnetic field set up by the stator windings. In other words, the speed of the rotor is now equal to the synchronous speed.

Remember that a synchronous motor must always be started as a three-phase AC induction motor with the DC field excitation disconnected. The DC field circuit is added only after the rotor accelerates to a value near the synchronous speed. The motor then operates as a synchronous motor, locked in step with the stator rotating field.

If an attempt is made to start a three-phase synchronous motor by first energizing the DC field circuit and then applying the three-phase voltage to the stator windings, the motor will not start because the net torque is zero. At the instant the three-phase voltage is applied to the stator windings, the magnetic field set up by the stator current turns at the synchronous speed. The rotor, with its magnetic poles of fixed polarity, is attracted first by an unlike stator pole and attempts to turn in that direction. However, before the rotor can turn, another stator pole of opposite polarity moves into position, and the rotor then attempts to turn in the opposite direction. Because of this action of the alternating poles, the net torque is zero and the motor does not start.

DC Field Excitation

In early models, the field circuit is excited from an external DC source. A DC generator may be coupled to the motor shaft to supply the DC excitation current.

Figure 17–5 shows the connections for a synchronous motor. A field rheostat in the separately excited field circuit varies the current in the field circuit. Changes in the field current affect the strength of the magnetic field developed by the revolving rotor. Small variations in the rotor field strength do not immediately affect the motor, which continues to operate at a constant speed. However, changes in the DC field excitation do change the power factor of a synchronous motor.

Brushless Solid-State Excitation

An improvement in synchronous motor excitation is the development of the brushless DC exciter. The commutator of a conventional direct-connected exciter is replaced with a three-phase, bridge-type, solid-state rectifier. The DC output is then fed directly to the motor field winding. Simplified circuitry is shown in Figure 17–6. A stationary field ring for the AC exciter receives DC from a small rectifier in the motor control cabinet. This rectifier is powered from the AC source. The exciter DC field is also adjustable. Rectifier solid-state diodes change the exciter AC output to DC. This DC is the source of excitation for the rotor field poles. Silicon-controlled rectifiers, activated by the solid-state field control circuit, replace electromechanical relays and the contactors of the conventional brush-type synchronous motor.

The field discharge resistor is inserted during motor starting. At motor synchronizing pull-in speed, the field discharge circuit is automatically opened and DC excitation is applied

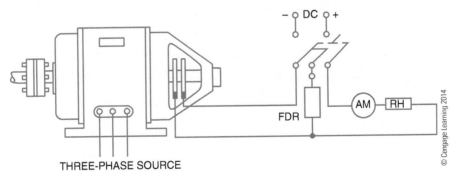

FIGURE 17–5 External connections for a synchronous motor.

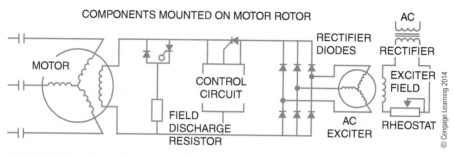

FIGURE 17-6 Simplified circuit for a brushless synchronous motor.

to the rotor field pole windings. Excitation is automatically removed if the motor pulls out of step (synchronization) due to an overload or a voltage failure. The brushless rotor is shown in Figure 17–7. Mounted on the rotor shaft is the armature of the AC exciter, the AC output of which is rectified to DC by the silicon diodes. Brush and commutator problems are eliminated with this system. (The stator of a brushless motor is similar to that of a brush type of motor.)

Power Factor

A poor lagging power factor results when the field current is decreased below normal by inserting all the resistance of the rheostat in the field circuit. The three-phase AC circuit to the stator supplies some magnetizing current, which helps strengthen the weak DC field. This magnetizing component of current lags the voltage by 90 electrical degrees. Because the magnetizing component of current becomes a large part of the total current input, a low lagging power factor results.

If a weak DC field is strengthened, the power factor improves. As a result, the three-phase AC circuit to the stator supplies less magnetizing current. The magnetizing component of current

FIGURE 17-7 Rotor of a brushless synchronous motor.

becomes a smaller part of the total current input to the stator winding, and the power factor increases. If the field strength is increased sufficiently, the power factor increases to unity, or 100%. When a power factor value of unity is reached, the three-phase AC circuit does not supply any rotor current and the DC field circuit supplies all current necessary to maintain a strong rotor field. The value of DC field excitation required to achieve unity power factor is called *normal field excitation*.

If the magnetic field of the rotor is strengthened further by increasing the DC field current above the normal field excitation value, the power factor decreases. However, the power factor is leading when the DC field is overexcited. The three-phase AC circuit feeding the stator winding delivers a demagnetizing component of current that opposes the too-strong rotor field. This action results in a weakening of the rotor field to its normal magnetic strength.

The diagrams in Figure 17–8 show how the DC field is aided or opposed by the magnetic field set up by the AC windings. It is assumed in Figure 17–8 that the DC field is stationary and

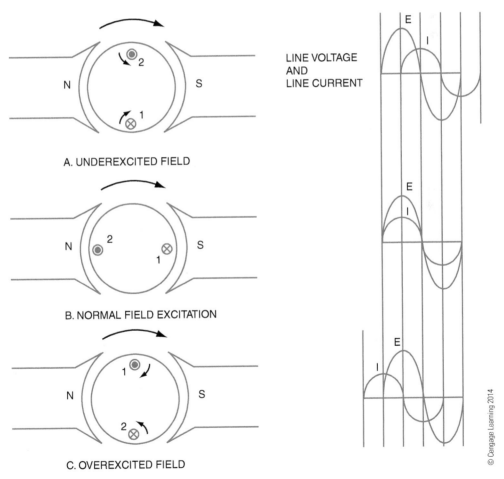

A. UNDEREXCITED FIELD

B. NORMAL FIELD EXCITATION

C. OVEREXCITED FIELD

LINE VOLTAGE AND LINE CURRENT

© Cengage Learning 2014

FIGURE 17–8 Field excitation in a synchronous motor.

a revolving armature is connected to the AC source. Keep in mind the fact that most synchronous motors have stationary AC windings and a revolving DC field. For either case, however, the principle of operation is the same.

Figure 17–9 shows two characteristic operating curves for a three-phase synchronous motor. With normal full-field excitation, the power factor has a peak value of unity, or 100% and the AC stator current is at its lowest value. As the DC field current is decreased in value, the power factor decreases into the lag quadrant, and there is a resulting rapid rise in the AC stator current. If the DC field current is increased above the normal field excitation value, the power factor decreases in the lead quadrant, and a rapid rise in the AC stator current results.

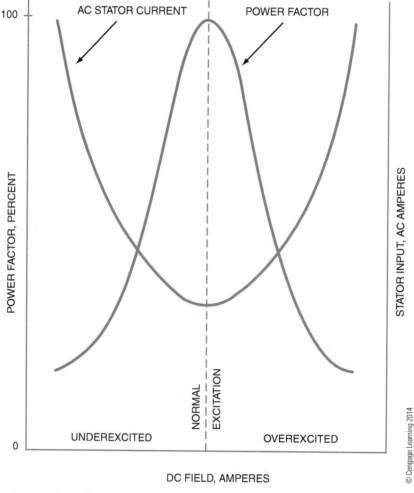

FIGURE 17–9 Characteristic operating curves for synchronous motors.

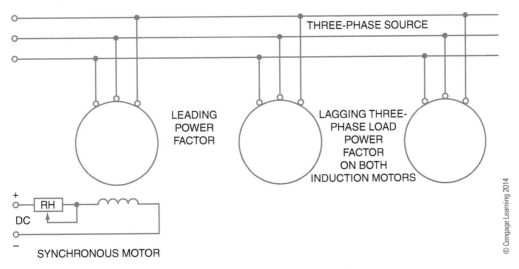

FIGURE 17–10 Synchronous motor used to correct power factor.

It has been shown that a synchronous motor operated with an overexcited DC field has a leading power factor. For this reason, a three-phase synchronous motor is often connected to a three-phase industrial feeder circuit with a low lagging power factor. In other words, the synchronous motor with an overexcited DC field helps correct the power factor of the industrial feeder circuit.

In Figure 17–10, two induction motors with lagging power factors are connected to an industrial feeder circuit. The synchronous motor connected to the same feeder is operated with an overexcited DC field. Because the synchronous motor can be adjusted so that the resulting power factor is leading, the power factor of the industrial feeder can be corrected until it reaches a value near unity, or 100%.

Reversing Rotation

The direction of rotation of a synchronous motor is reversed by interchanging any two of the three line leads feeding the stator winding. The direction of rotation of the motor does not change if the two conductors of the DC source are interchanged.

INDUSTRIAL APPLICATIONS

The three-phase synchronous motor is used when a prime mover having a constant speed from a no-load condition to full load is required, such as fans, air compressors, and pumps. The synchronous motor is used in some industrial applications to drive a mechanical load and also to correct the power factor. In some applications, this type of motor is used only to correct the power factor of an industrial power system. When the synchronous motor is used only to correct the power factor and does not drive any mechanical load, it serves the same purpose as a bank of capacitors used for power factor correction. Therefore, in such an installation the motor is called a *synchronous capacitor*.

Three-phase synchronous motors up to a rating of 10 hp are usually started directly across the rated three-phase voltage. Synchronous motors of larger sizes are started through a starting compensator or an automatic starter. In this type of starting, the voltage applied to the motor terminals at the instant of start is about half the value of the rated line voltage, and the starting surge of current is limited.

SUMMARY

The AC synchronous motor is used where speed must be kept constant. As the name implies, the motor runs at the designed synchronous speed. The principle used in the larger three-phase synchronous motors is to provide a DC field for the rotor. The methods may vary in the application of the DC. Some motors use an external DC source and feed the DC to the rotor via slip rings. Other motors control a magnetic field to the rotor and use solid-state rectifiers to create DC in the rotor. In either case, the rotor field can change the power factor of the synchronous motor and allow it to act as a source of leading-power factor, thereby correcting the normal lagging power factor of an industrial power system.

ACHIEVEMENT REVIEW

A. Fill in the answers to questions 1 through 9.

1. List the basic parts of a three-phase synchronous motor.

_____

_____

_____

_____

_____

2. What is an amortisseur winding?

_____

_____

_____

_____

_____

3. Explain the proper procedure to use in starting a synchronous motor.

_____

_____

_____

_____

_____

4. A three-phase synchronous motor with six stator poles and six rotor poles is operated from a three-phase, 60 Hz line of the correct voltage rating. Determine the speed of the motor._____

_____

_____

_____

_____

_____

_____

5. How is a leading power factor obtained with a three-phase synchronous motor?

_____

_____

_____

_____

_____

6. What is the purpose of a rheostat in the separately excited DC field circuit of a synchronous motor?

_____

_____

_____

_____

_____

7. How is the direction of rotation of a three-phase synchronous motor reversed?

_____

_____

_____

_____

_____

8. State two important applications for three-phase synchronous motors.

_____

_____

9. What is a synchronous capacitor? _____

_____

_____

_____

B. Select the correct answer for the statements in items 10 through 14, and place the corresponding letter in the space provided. _____

10. A synchronous motor must be started _____
    a. with full DC in the field circuit.
    b. with weak DC in the field circuit.
    c. as an induction motor.
    d. when the power factor is low.

11. The speed of a synchronous motor _____
    a. is constant from no load to full load.
    b. drops from no load to full load.
    c. increases from no load to full load.
    d. is variable from no load to full load.

12. A synchronous motor with an underexcited DC field has _____
    a. a leading power factor.
    b. a lagging power factor.
    c. less synchronous speed.
    d. no effect.

13. The power factor of a synchronous motor can be varied by changing the _____
    a. brush polarity.
    b. phase rotation.
    c. speed of rotation.
    d. field excitation.

14. A synchronous motor running on the three-phase line voltage serves the same function of power factor correction as _____
    a. a bank of resistors.
    b. a bank of capacitors.
    c. an induction motor.
    d. a wound-rotor motor.

UNIT

18

SUMMARY REVIEW OF UNITS 12–17

OBJECTIVE

- To provide the student with an opportunity to evaluate the knowledge and understanding acquired in the study of the previous six units.

1. List three types of three-phase AC motors.

   a. _____

   b. _____

   c. _____

2. Insert the word or phrase to complete each of the following statements.

   a. The speed of a three-phase induction motor falls slightly from a no-load condition to a full load. This is true of a three-phase induction motor with a _____ rotor.

   b. A speed controller is used only with a three-phase induction motor of the _____ type.

   c. When all resistance of the speed controller is inserted in the secondary circuit of a three-phase _____ _____ induction motor, the starting torque is very good.

   d. A three-phase _____ motor is operated with an overexcited DC field to obtain a leading power factor.

   e. The speed of a three-phase _____ motor remains constant from a no-load condition to full load if the operating frequency remains constant.

3. State two advantages of using a squirrel-cage induction motor.

   _____

   _____

   _____

   _____

4. State one disadvantage of using a squirrel-cage induction motor. _____

   _____

   _____

   _____

5. Explain how the direction of rotation of a three-phase AC induction motor is reversed.

   _____

   _____

   _____

   _____

   _____

6. A two-pole, 60 Hz, three-phase AC induction motor has a full-load speed of 3475 RPM. Determine the synchronous speed of this motor. _____

_____

_____

_____

_____

7. Determine the percent slip of the motor in question 6. _____

_____

_____

_____

8. What is the purpose of starting protection for a three-phase motor? _____

_____

_____

_____

9. What is the purpose of running protection for a three-phase motor? _____

_____

_____

_____

10. Show the connection diagram for the nine terminal leads of a wye-connected, three-phase motor rated at 230/460 volts for three-phase operation on 230 volts.

11. Explain how the running overload protection for a three-phase motor rated at more than 1 hp would be selected. _____

_____

_____

_____

_____

12. Why are electronic drives used with large three-phase AC induction motors? _____

_____

_____

_____

_____

13. Why is a wound-rotor induction motor used in place of a squirrel-cage induction motor for some industrial applications? _____

_____

_____

_____

_____

_____

14. Explain how the direction of rotation of a wound-rotor induction motor can be reversed.

_____

_____

_____

_____

_____

15. Insert the correct word or phrase to complete each of the following statements.

    a. The speed of a wound-rotor induction motor is _____ by inserting resistance in the rotor circuit through a speed controller.

  b. The starting surge of current of a wound-rotor induction motor is minimized by

     _____.

  c. The rotation of a wound-rotor induction motor is _____ by changing any two of the three leads feeding from the rotor slip rings to the speed controller.

  d. The _____ of a wound-rotor induction motor is very good if all the resistance of the speed controller is inserted in the rotor circuit.

  e. The efficiency of a wound-rotor induction motor operating at rated load with all the resistance inserted in the rotor circuit is _____.

16. Draw a schematic connection diagram of a wound-rotor induction motor that is started by means of an ATL magnetic motor starter controlled from a pushbutton station.

17. Explain how the direction of rotation of a three-phase synchronous motor is reversed.

    _____

    _____

    _____

    _____

18. List two important applications for three-phase synchronous motors.

    _____

    _____

19. A three-phase synchronous motor with four stator poles and four rotor poles is operated from a three-phase, 60 Hz line of the correct voltage rating. Determine the speed of the motor. _____

    _____

    _____

    _____

    _____

20. Explain the correct procedure for starting a three-phase synchronous motor.

_____

_____

_____

_____

_____

21. Insert the correct word or phrase to complete each of the following statements.

a. The speed of a synchronous motor is _____ from no load to full load.

b. A synchronous motor with an underexcited DC field has a _____ power factor.

c. A three-phase _____ motor must be started as an induction motor.

22. Explain what is meant by the term *jogging*. _____

_____

_____

23. Explain what is meant by the term *plugging*. _____

_____

_____

24. What identifying information should appear on a motor controller to comply with the requirements of the *National Electrical Code*®?

_____

_____

_____

_____

_____

25. What is meant by a PWM AC drive?

_____

_____

_____

26. Insert the correct word or phrase to complete each of the following statements.
   a. A controller with _____ may be used to stop a motor quickly.
   b. When fuses are used as protection for a three-phase, three-wire, ungrounded branch-motor circuit, the fuses must be installed in _____ line leads.
   c. Motors operating on a three-phase, three-wire ungrounded system require _____ thermal overload units for running overcurrent protection.

27. How is dynamic braking applied to an induction motor?

_____

_____

_____

_____

_____

28. How is dynamic braking applied to a synchronous motor?

_____

_____

_____

29. Draw a schematic diagram of an ATL magnetic switch connected to a three-phase AC induction motor. The magnetic switch has jogging capability. Include in the connection diagram the main relay coil; the pushbutton station with start, jog, and stop pushbuttons; and the maintaining contact.

30. Draw a schematic diagram of a wye–delta, open-transition starter, complete with push-button station, connected to a three-phase motor.

31. On periodic tests, a motor winding suddenly drops to a low resistance value. Testing with a _____ determines this condition.

32. For the sleeve bearings of an AC motor, explain how the old oil is removed and the bearings cleaned and lubricated.

_____

_____

_____

_____

33. Place the correct answers in each of the spaces provided in Figure 18–1. Refer to the *National Electrical Code*®.

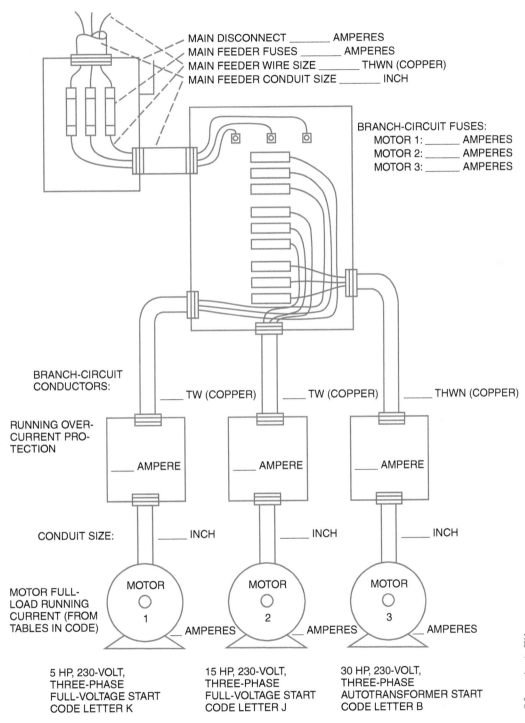

MAIN DISCONNECT _____ AMPERES
MAIN FEEDER FUSES _____ AMPERES
MAIN FEEDER WIRE SIZE _____ THWN (COPPER)
MAIN FEEDER CONDUIT SIZE _____ INCH

BRANCH-CIRCUIT FUSES:
    MOTOR 1: _____ AMPERES
    MOTOR 2: _____ AMPERES
    MOTOR 3: _____ AMPERES

BRANCH-CIRCUIT
CONDUCTORS:

_____ TW (COPPER)   _____ TW (COPPER)   _____ THWN (COPPER)

RUNNING OVER-
CURRENT PRO-
TECTION

_____ AMPERE   _____ AMPERE   _____ AMPERE

CONDUIT SIZE: _____ INCH   _____ INCH   _____ INCH

MOTOR
1

MOTOR
2

MOTOR
3

MOTOR FULL-
LOAD RUNNING
CURRENT (FROM
TABLES IN CODE)

_____ AMPERES   _____ AMPERES   _____ AMPERES

5 HP, 230-VOLT,   15 HP, 230-VOLT,   30 HP, 230-VOLT,
THREE-PHASE   THREE-PHASE   THREE-PHASE
FULL-VOLTAGE START FULL-VOLTAGE START AUTOTRANSFORMER START
CODE LETTER K   CODE LETTER J   CODE LETTER B

© Cengage Learning 2014

FIGURE 18–1 Fill in the blanks for question 33.

SINGLE-PHASE INDUCTION MOTORS

OBJECTIVES

After studying this unit, the student should be able to

- describe the basic operation of the following types of induction motors:

- split-phase motor (both single and dual voltage).

- capacitor-start, induction-run motor (both single and dual voltage).

- capacitor-start, capacitor-run motor with one capacitor.

- capacitor-start, capacitor-run motor with two capacitors.

- capacitor-start, capacitor-run motor having an auto-transformer with one capacitor.

- compare the motors in the preceding listing with regard to starting torque, speed performance, and power factor at the rated load.

- identify shaded pole motor components and operation.

The two principal types of single-phase induction motors are the split-phase motor and the capacitor motor. Both types of single-phase induction motors usually have a fractional horsepower rating. The split-phase motor is used to operate such devices as washing machines, small water pumps, oil burners, and other types of small loads not requiring a strong starting torque. The capacitor motor generally is used with devices requiring a strong starting torque, such as refrigerators and compressors. Both types of single-phase induction motors are relatively low in cost, have a rugged construction, and exhibit a good operating performance.

CONSTRUCTION OF A SPLIT-PHASE INDUCTION MOTOR

The split-phase induction motor basically consists of a stator, a rotor, a centrifugal switch located inside the motor, two end shields housing the bearings that support the rotor shaft, and a cast steel frame into which the stator core is pressed. The two end shields are bolted to the cast steel frame. The bearings housed in the end shields keep the rotor centered within the stator so that it will rotate with a minimum of friction and without striking or rubbing the stator core.

The stator for a split-phase motor consists of two windings held in place in the slots of a laminated steel core. The two windings consist of insulated coils distributed and connected to make up two windings spaced 90 electrical degrees apart. One winding is the running winding, and the second winding is the starting winding.

The running winding consists of insulated copper wire placed at the bottom of the stator slots. The wire size in the starting winding is smaller than that of the running winding. These coils are placed on top of the running winding coils in the stator slots nearest the rotor.

Both the starting and running windings are connected in parallel to the single-phase line when the motor is started. After the motor accelerates to a speed equal to approximately two-thirds to three-quarters of the rated speed, the starting winding is disconnected automatically from the line by means of a centrifugal switch.

The rotor for the split-phase motor has the same construction as that of a three-phase AC induction motor; that is, the rotor consists of a cylindrical core assembled from steel laminations. Copper bars are mounted near the surface of the rotor. The bars are brazed or welded to two copper end rings. In some motors, the rotor is a one-piece cast aluminum unit.

Figure 19–1 shows a typical squirrel-cage rotor for a single-phase induction motor. This type of rotor requires little maintenance because there are no windings, brushes, slip rings, or commutators. Note in the figure that the rotor fan is part of the squirrel-cage rotor assembly. These rotor fans maintain air circulation through the motor to prevent a large increase in the temperature of the windings.

The centrifugal switch is mounted inside the motor. The centrifugal switch disconnects the starting winding after the rotor reaches a predetermined speed, usually two-thirds to three-quarters of the rated speed. The switch consists of a stationary part and a rotating part. The stationary part is mounted on one of the end shields and has two contacts that act like a single-pole, single-throw switch. The rotating part of the centrifugal switch is mounted on the rotor.

FIGURE 19–1 Cast aluminum squirrel-cage rotor.

A simple diagram of the operation of a centrifugal switch is shown in Figure 19–2. When the rotor is at a standstill, the pressure of the spring on the fiber ring of the rotating part keeps the contacts closed. When the rotor reaches approximately three-quarters of its rated speed, the centrifugal action of the rotor causes the spring to release its pressure on the fiber ring and the contacts to open. As a result, the starting winding circuit is disconnected from the line. Figure 19–3 shows a typical centrifugal switch used with split-phase induction motors.

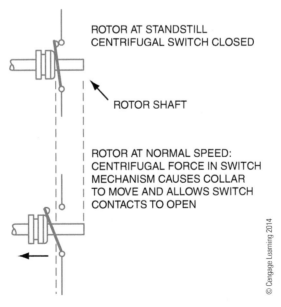

ROTOR AT STANDSTILL
CENTRIFUGAL SWITCH CLOSED

ROTOR SHAFT

ROTOR AT NORMAL SPEED:
CENTRIFUGAL FORCE IN SWITCH
MECHANISM CAUSES COLLAR
TO MOVE AND ALLOWS SWITCH
CONTACTS TO OPEN

FIGURE 19–2 The operation of a centrifugal switch.

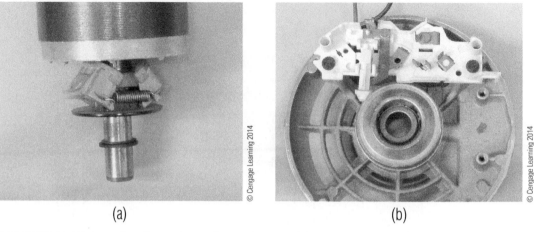

(a)   (b)

FIGURE 19–3 (A) Centrifugal switch weight mechanism. (B) Stationary part of centrifugal switch.

Principle of Operation

When the circuit to the split-phase induction motor is closed, both the starting and running windings are energized in parallel. Because the running winding consists of a relatively large size of wire, its resistance is low. Recall that the running winding is placed at the bottom of the slots of the stator core. As a result, the inductive reactance of this winding is comparatively high due to the mass of iron surrounding it. Because the running winding has a low resistance and a high inductive reactance, the current of the running winding lags behind the voltage approximately 90 electrical degrees.

The starting winding consists of a smaller size of wire; therefore, its resistance is high. Because the winding is placed near the top of the stator slots, the mass of iron surrounding it is comparatively small and the inductive reactance is low. Therefore, the starting winding has a high resistance and a low inductive reactance. As a result, the current of the starting winding is nearly in phase with the voltage.

The two windings create the split in the single phase of incoming voltage by the way the current reacts to the two windings (see Figure 19–4). Note that the phase split between the currents in the starting winding and the running winding is not the ideal 90°, but the effect is the same as splitting the magnetic effects of the coils to produce the rotating magnetic field needed to get the rotor spinning. After the starting winding is disconnected, the rotor continues to spin because of the pulsating magnetic field of the running winding and the momentum of the spinning rotor and load.

The current of the running winding lags the current of the starting winding by approximately 30 electrical degrees. These two currents, spaced 30 electrical degrees apart, pass through these windings, and a rotating magnetic field is developed. This field travels around

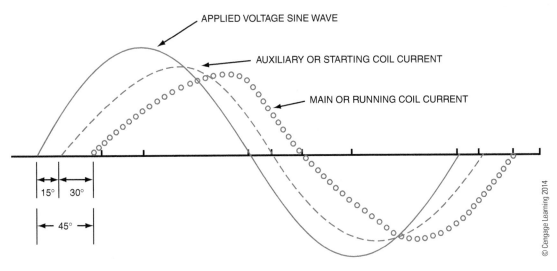

FIGURE 19–4 Sinewaves of split-phase motor voltage and current characteristics.

the inside of the stator core. The speed of the magnetic field is determined using the same procedure given for a three-phase induction motor.

If a split-phase induction motor has four poles on the stator windings and is connected to a single-phase, 60-Hz source, the synchronous speed of the revolving field is

$$\text{Synchronous RPM} = \frac{120 \times f}{4} \qquad f = \text{Frequency in hertz}$$

$$\text{RPM} = \frac{120 \times 60}{4}$$

$$\text{RPM} = 1800$$

As the rotating stator field travels at the synchronous speed, it cuts the copper bars of the rotor and induces voltages in the bars of the squirrel-cage winding. These induced voltages set up currents in the rotor bars. As a result, a rotor field is created that reacts with the stator field to develop the torque that causes the rotor to turn.

As the rotor accelerates to the rated speed, the centrifugal switch disconnects the starting winding from the line. The motor then continues to operate using only the running winding. Figure 19–5 illustrates the connections of the centrifugal switch at the instant the motor starts (switch closed) and when the motor reaches its normal running speed (switch open).

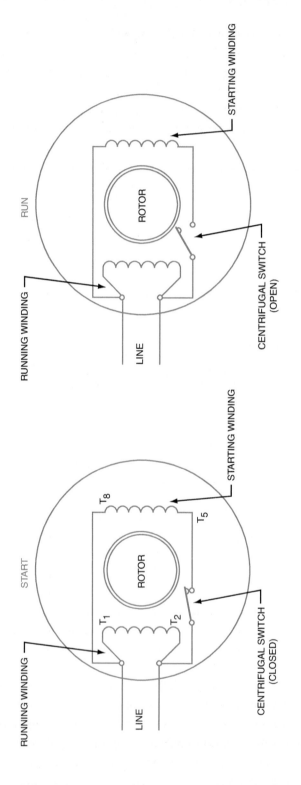

SPLIT-PHASE INDUCTION MOTOR

RUN

RUNNING WINDING

STARTING WINDING

ROTOR

LINE

CENTRIFUGAL SWITCH
(OPEN)

START

RUNNING WINDING

STARTING WINDING

ROTOR

T_8

T_5

T_1

T_2

LINE

CENTRIFUGAL SWITCH
(CLOSED)

THE CENTRIFUGAL SWITCH OPENS AT APPROXIMATELY 75% OF RATED SPEED.

THE STARTING WINDING HAS HIGH RESISTANCE AND LOW INDUCTIVE REACTANCE.
THE RUNNING WINDING HAS LOW RESISTANCE AND HIGH INDUCTIVE REACTANCE.
(PRODUCES 45° TO 60° PHASE ANGLE FOR STARTING TORQUE.)

© Cengage Learning 2014

FIGURE 19–5 Connections of the centrifugal switch at start and at run.

A split-phase motor must have both the starting and running windings energized when the motor is started. The motor resembles a two-phase induction motor in which the currents of these two windings are approximately 90 electrical degrees out of phase. The voltage source, however, is single phase; therefore, the motor is called a split-phase motor because it starts like a two-phase motor from a split single-phase line. After the motor accelerates to a value near its rated speed, it operates on the running winding as a single-phase induction motor.

If the centrifugal switch contacts fail to close when the motor stops, then the starting winding circuit is still open. When the motor circuit is re-energized, the motor will not start. The motor must have both the starting and running windings energized at the instant the motor circuit is closed to create the necessary starting torque. If the motor does not start but simply gives a low humming sound, then the starting winding circuit is open. Either the centrifugal switch contacts are not closed, or there is a break in the coils of the starting windings. *This is an unsafe condition.* The running winding will draw excessive current and, therefore, the motor must be removed from the line supply.

If the mechanical load is too great when a split-phase motor is started, or if the terminal voltage applied to the motor is low, then the motor may fail to reach the speed required to operate the centrifugal switch.

The starting winding is designed to operate at-line voltage for a period of only 3 or 4 seconds while the motor is accelerating to its rated speed. It is important that the starting winding be disconnected from the line by the centrifugal switch as soon as the motor accelerates to 75% of the rated speed. Operation of the motor on its starting winding for more than 60 seconds may burn the insulation on the winding or cause the winding to burn out.

Some motors do not contain the mechanical type of starting switch. Instead they use a switch external to the motor. There are three general types of external starting switches in use that are external to the motor. These starting winding switches may need to be located remotely from the motor for ease of maintenance or because the switching may create an objectionable arc in a hazardous location. In these cases, a voltage- or current-operated switch may be used for mechanical switching, or solid-state relays may be used for arcless switching.

The voltage-operated switch is shown in Figure 19–6. Voltage-operated switches are connected across the starting winding. These switches are used with capacitor-start, single-phase motors, but will be explained here with other starting switches. As the motor power is applied, current is allowed to flow through the normally closed switch to the starting winding and the running winding. The starting winding is a low-impedance load because of the starting capacitor in the circuit. The large inrush current causes a large voltage drop across the series capacitor and little across the motor coil. Very little voltage is dropped across the starting winding, and therefore little voltage is dropped across the coil for the starting switch, which keeps the relay contacts closed. As the rotor begins to spin, the CEMF generated into the starting coil increases, and the coil begins to drop a greater percentage of the line voltage. Now the starter relay coil energizes and opens the contacts, interrupting current to the starting winding. The induced voltage from the spinning rotor is enough to hold the relay contacts open while the motor is spinning.

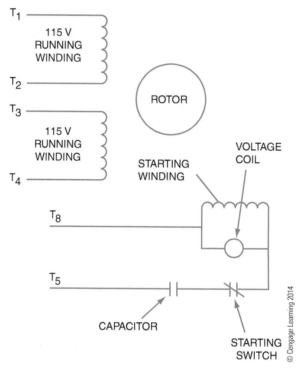

FIGURE 19–6 Single-phase motor using a voltage-operated coil for starting winding circuit for a capacitor-start motor.

A second style of mechanical switch is the current-operated starting switch shown in Figure 19–7. The current coil is connected in series with the running winding, as shown in Figure 19–8. As power is applied to the single-phase motor, the current inrush to the running winding is relatively large (six to ten times running current). This large inrush current causes the current relay to pull the contacts closed, which are in series with the starting winding. These

FIGURE 19–7 Current type of starting relay.

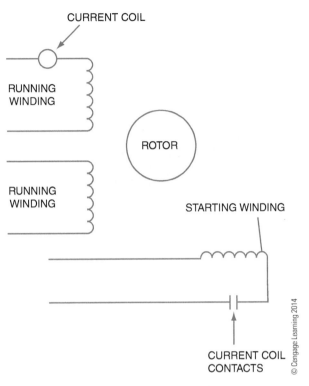

CURRENT COIL

RUNNING
WINDING

RUNNING
WINDING

ROTOR

STARTING WINDING

CURRENT COIL
CONTACTS

© Cengage Learning 2014

FIGURE 19–8 Single-phase motor with current-operated coil for starting winding circuit.

contacts allow current to flow to the starting winding, and the rotor begins to spin. As the motor increases in speed, the inrush current diminishes and the current coil contacts are allowed to open. The starting winding is now disconnected from the line power, and the motor operates as a normal single-phase motor.

A third type of starting relay is the electronic relay, shown in Figure 19–9. This is used effectively for arcless switching of the starting winding. The relay is connected as shown in Figure 19–10. The concept is the same as the current-operated relay, but the sensing device is a thermistor. This thermistor is a solid-state device that will react to a change in temperature by increasing its resistance (positive temperature coefficient). Because the thermistor is in series with the starting winding, the current to the starting winding causes the temperature of the thermistor to increase, and the resistance also begins to increase from 3 to 4 ohms up to a few thousand ohms. This has the same effect as opening the switch to the winding. A small amount of current is still flowing through the high resistance to the starting winding. This current does not affect the motor operation, but is needed to keep the thermistor warm and the resistance high. This type of starting switch is sensitive to temperature, and if it is not allowed to cool sufficiently, it may prevent the starting winding from performing normally during starting. Be sure to allow the motor to cool down between successive starts.

FIGURE 19-9　Solid-state running relay.

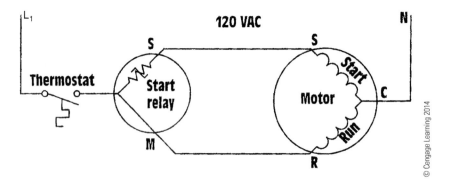

FIGURE 19-10　Solid-state starting relay connection.

To reverse the rotation of the motor, simply interchange the leads of the starting winding (Figure 19–11). This causes the direction of the field set up by the stator windings to become reversed. As a result, the direction of rotation is reversed. The direction of rotation of the split-phase motor can also be reversed by interchanging the two running winding leads. Normally, the starting winding is used for reversing.

Single-phase motors often have dual-voltage ratings of 115 volts and 230 volts. To obtain these ratings, the running winding consists of two sections. Each section of the winding is rated at 115 volts. One section of the running winding is generally marked T_1 and T_2, and the other section is marked T_3 and T_4. If the motor is to be operated on 230 volts, the two 115-volt windings are connected in series across the 230-volt line. If the motor is to be operated on 115 volts, then the two 115-volt windings are connected in parallel across the 115-volt line.

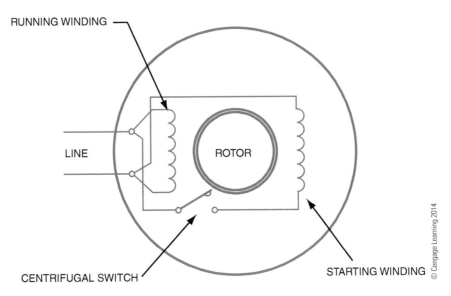

FIGURE 19-11 Reversing direction of rotation on split-phase induction motor usually means that the connections to the starting winding is reversed.

The starting winding usually consists of only one 115-volt winding. The leads of the starting winding are generally marked T_5 and T_8. If the motor is to be operated on 115 volts, both sections of the running winding are connected in parallel with the starting winding (Figure 19–12).

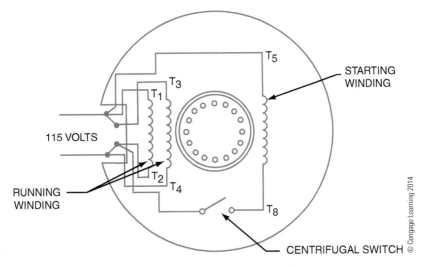

FIGURE 19-12 Dual-voltage motor connected for 115 volts.

For 230-volt operation, the connection jumpers are changed in the terminal box so that the two 115-volt sections of the running winding are connected in series across the 230-volt line (Figure 19–13). Note that the 115-volt starting winding is connected in parallel with one section of the running winding. The voltage drop across this section of the running winding is 115 volts, and the voltage across the starting winding is also 115 volts.

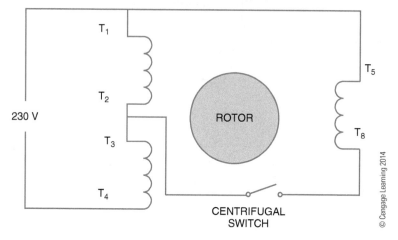

FIGURE 19–13 Dual-voltage motor connected for 230 volts.

Some dual-voltage, split-phase motors have a starting winding with two sections as well as a running winding with two sections. The running winding sections are marked T_1 and T_2 for one section and T_3 and T_4 for the other section. One section of the starting winding is marked T_5 and T_6, and the second section of the starting winding is marked T_7 and T_8.

The National Electrical Manufacturers Association (NEMA) has color coded the terminal leads. If colors are used, they should be coded as follows: T_1, blue; T_2, white; T_3, orange; T_4, yellow; T_5, black; and T_8, red. If there are leads with T_6 and T_7, there are no assigned colors from the standard color markings.

Figure 19–14 shows the winding arrangement for a dual-voltage motor with two starting windings and two running windings. The correct connections for 115-volt operation and for 230-volt operation are given in the table shown in Figure 19–14.

The speed regulation of a split-phase induction motor is very good. It has a speed performance from no load to full load that is similar to that of a three-phase AC induction motor. The percent slip on most fractional horsepower split-phase motors is from 4% to 6%.

The starting torque of the split-phase motor is comparatively poor. The low resistance and high inductive reactance in the running winding circuit and the high resistance and low inductive reactance in the starting winding circuit cause the two current values to be considerably less than 90 electrical degrees apart. The currents of the starting and running windings in many split-phase motors are only 30 electrical degrees out of phase with each other. As a result, the field set up by these currents does not develop a strong starting torque.

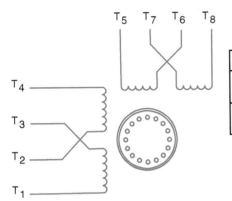

VOLTAGE RATING	L_1	L_2	TIE TOGETHER
115 VOLTS	$T_1, T_3,$ T_5, T_7	$T_2, T_4,$ T_6, T_8	——————
230 VOLTS	T_1, T_5	T_4, T_8	T_2 AND $T_3,$ T_6 AND T_7

© Cengage Learning 2014

FIGURE 19–14 Winding arrangements for dual-voltage motor with two starting and two running windings.

CAPACITOR-START, INDUCTION-RUN MOTOR

The construction of a capacitor-start motor is nearly the same as that of a split-phase induction motor. For the capacitor-start motor, however, a capacitor is connected in series with the starting windings. The capacitor is usually mounted in a metal casing on top of the motor. The capacitor may be mounted in any convenient external position on the motor frame and, in some cases, may be mounted inside the motor housing. The capacitor provides a higher starting torque than is obtainable with the standard split-phase motor. In addition, the capacitor limits the starting surge of current to a lower value than is developed by the standard split-phase motor.

The capacitor-start induction motor is used on refrigeration units, compressors, oil burners, and for small machine equipment, as well as for other applications that require a strong starting torque.

Principle of Operation

When the capacitor-start motor is connected for lower voltage and is started, both the running and starting windings are connected in parallel across the line voltage and the centrifugal switch is closed. The starting winding, however, is connected in series with the capacitor. When the motor reaches a value of approximately 75% of its rated speed, the centrifugal switch opens and disconnects the starting winding and the capacitor from the line. The motor then operates as a single-phase induction motor using only the running winding. The capacitor is used to improve the starting torque and does not improve the power factor of the motor.

To produce the necessary starting torque, the revolving magnetic field must have a better split of the single-phase line current. The capacitor added in series with the starting winding causes the current in the starting winding to be shifted to a leading current ahead of the voltage, instead of a lagging current as in the split-phase motor. Figure 19–15 illustrates the leading effect caused by the addition of a starting rated capacitor. The current in the starting winding

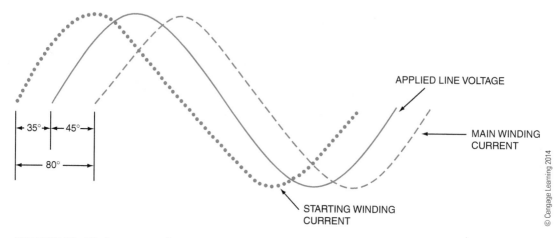

FIGURE 19–15 Sinewave of capacitor-start motor voltage and current characteristics.

leads the line voltage by approximately 35°, and the current in the running winding lags the line voltage by 45°. This differential creates a split of 80°, nearing the optimum of 90 electrical degrees. This creates a better orientation of the magnetic fields of the stator and therefore creates more torque than a similarly sized split-phase motor.

Defective capacitors are a frequent cause of malfunctions in capacitor-start, induction-run motors. Some capacitor failures that can occur are listed here:

- The capacitor may short itself out, as evidenced by a lower starting torque.
- The capacitor may be "opened," in which case the starting winding circuits will be open, causing the motor to fail to start.
- The capacitor may short-circuit and cause the fuse protection for the branch motor circuit to blow. If the fuse ratings are high and do not interrupt the power supply to the motor soon enough, the starting winding may burn out.
- Starting capacitors can short-circuit if the motor is turned on and off many times in a short period of time. To prevent capacitor failures, many motor manufacturers recommend that a capacitor-start motor be started no more than 20 times per hour. Therefore, this type of motor is used only in applications where there are relatively few starts in a short time period.

The speed performance of a capacitor-start motor is very good. The change in percent slip from a no-load condition to full load is from 4% to 6%. The speed performance then is the same as that of a standard split-phase motor.

The leads of the starting winding circuit are interchanged to reverse the direction of rotation of a capacitor-start motor. As a result, the direction of rotation of the magnetic field developed by stator windings reverses in the stator core, and the rotation of the rotor is reversed (see Figure 19–16 for reversing lead connections).

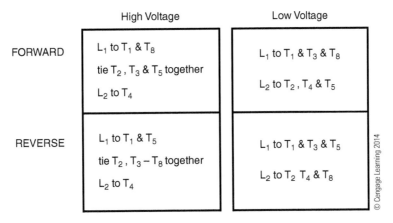

FIGURE 19–16 Chart for two running windings and one starting winding.

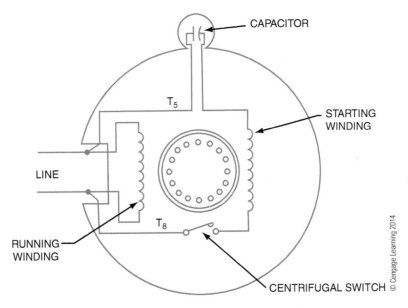

FIGURE 19–17 Connections for a capacitor-start induction motor.

Figure 19–17 shows a diagram of the circuit connections of a capacitor-start motor before the starting winding leads are interchanged to reverse the direction of rotation of the rotor. The diagram in Figure 19–18 shows the circuit connections of the motor after the starting winding leads are interchanged to reverse the direction of rotation.

A second method of changing the direction of rotation of a capacitor-start motor is to interchange the two running winding leads. However, this method is seldom used.

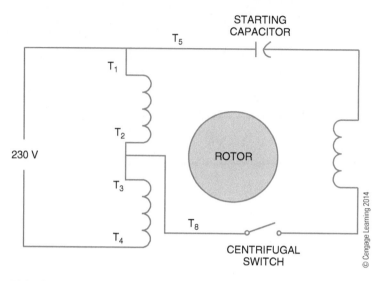

FIGURE 19–18 Connections for reversing a capacitor-start, induction-run motor.

Capacitor-start, induction-run motors often have dual-voltage ratings of 115 volts and 230 volts. The connections for a capacitor-start motor are the same as those for split-phase induction motors, in that the starting winding is connected across half of the series running winding.

CAPACITOR-START, CAPACITOR-RUN MOTOR OR TWO-VALUE CAPACITOR MOTOR

The capacitor-start, capacitor-run motor is similar to the capacitor-start, induction-run motor, except that the starting winding and capacitor are connected in the circuit at all times. This motor has a very good starting torque. The power factor at the rated load is nearly 100%, or unity, because that a capacitor is used in the motor at all times.

This type of motor can have several different designs. One type of capacitor-start, capacitor-run motor has two stator windings that are spaced 90 electrical degrees apart. The main or running winding is connected directly across the rated line voltage. A capacitor is connected in series with the starting winding, and this series combination is also connected across the rated line voltage. A centrifugal switch is not used, because the starting winding is energized through the entire operating period of the motor.

Figure 19–19 illustrates the internal connections for a capacitor-start, capacitor-run motor using one value of capacitance.

To reverse the rotation of this motor, the lead connections of the starting winding must be interchanged. This type of capacitor-start, capacitor-run motor is quiet in operation and is used on air compressors, fans, and small woodworking and metalworking machines.

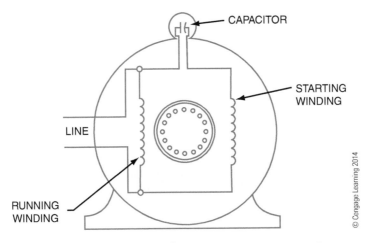

FIGURE 19–19 Connections for a capacitor-start, capacitor-run motor.

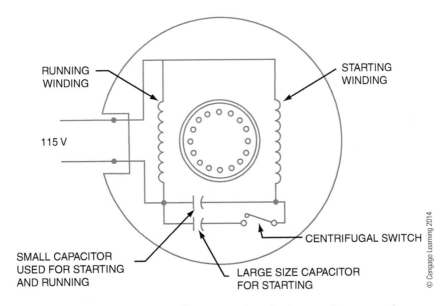

FIGURE 19–20 Connections for a capacitor-start, capacitor-run motor.

A second type of capacitor-start, capacitor-run motor has two capacitors. Figure 19–20 shows a diagram of the internal connections of the motor. At the instant the motor is started, the two capacitors are in parallel. When the motor reaches 75% of the rated speed, the centrifugal switch disconnects the larger capacity capacitor. The motor then operates with the smaller capacitor only connected in series with the starting winding.

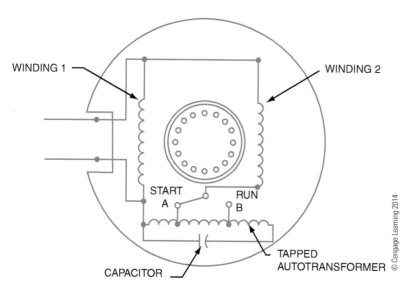

FIGURE 19–21 Connections for a capacitor-start, capacitor-run motor with autotransformer.

This type of motor has a very good starting torque, good speed regulation, and a power factor of nearly 100% at rated load. Applications for this type of motor include furnace stokers, refrigerator units, and compressors.

A third type of capacitor-start, capacitor-run motor has an autotransformer with one capacitor. This motor has a high starting torque and a high operating power factor. Figure 19–21 shows a diagram of the internal connections for this motor. When the motor is started, the centrifugal switch connects winding 2 to point A on the tapped autotransformer. As the capacitor is connected across the maximum transformer turns, it receives maximum voltage output on start. The capacitor thus is connected across a value of approximately 500 volts. As a result, there is a high value of leading current in winding 2, and a strong starting torque is developed.

When the motor reaches approximately 75% of the rated speed, the centrifugal switch disconnects the starting winding from point A and reconnects this winding to point B on the autotransformer. Less voltage is applied to the capacitor, but the motor operates with both windings energized. Thus, the capacitor maintains a power factor near unity at the rated load.

The starting torque of this motor is very good, and the speed regulation is satisfactory. Applications requiring these characteristics include large refrigerators and compressors.

A permanent-split capacitor motor is shown in Figure 19–22. In this type of motor, a low microfarad, oil-filled capacitor is connected in series with one of the identical coil windings.

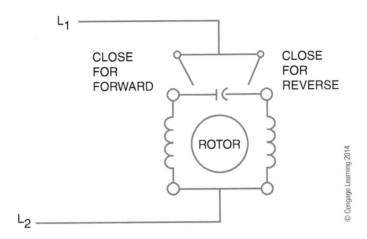

FIGURE 19-22 Permanent-split capacitor motor with forward and reverse switches shown.

By placing a capacitor in series with one winding, the current in that winding leads the current in the other winding, causing a split in the magnetic fields and causing the motor to rotate from the field with the capacitor toward the field without the capacitor. This motor is used for low torque requirements that may require frequent reversing. By simply closing one switch or the other, the capacitor will be in series with different windings, thus reversing the direction of motor rotation.

SHADED POLE MOTORS

Shaded pole motors are among the simplest and cheapest motors to construct. The principle of operation uses the effects of induction not only into the squirrel-cage rotor, but also into parts of the stator that create a rotating magnetic field from a single phase of input voltage. These motors are typically fractional horsepower ratings and are used in applications that do not require a great deal of starting torque.

In a simple unidirectional motor, a ring (shading ring) of solid conductor is short-circuited and embedded in one side of a stator winding (see Figure 19–23 and Figure 19–24).

As the voltage is applied to the top and bottom coils, the shading coil has voltage induced into it. Lenz's law states that the effect of induction always opposes the cause of induction. Therefore, the magnetic field developed by the shading ring as current flows through its shorted winding opposes the main flux. This causes the main magnetic field to be shifted away from the shading ring.

As the applied voltage waveform begins to decrease from its peak value, the magnetic lines of force also decrease. The effect on the shading ring is the opposite. As main current flow decreases,

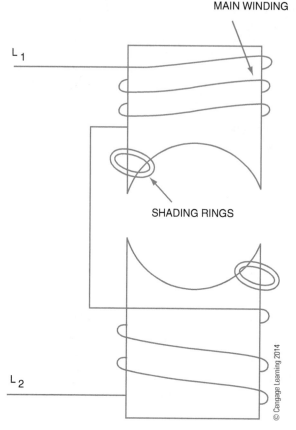

MAIN WINDING

L₁

SHADING RINGS

L₂

© Cengage Learning 2014

FIGURE 19-23 Schematic of shaded pole motor.

the magnetic effect of the shading ring tends to keep the same polarity as the main pole but increases in strength. This causes the stator pole to move from the main (unshaded) pole toward the shaded pole (see Figure 19–25).

Position 1 Top of stator: Main coil magnetic polarity is north pole. Shading ring opposes main with a South polarity. (Use Figure 19–23 as reference.)

Position 2 Top of stator: Main pole is less strong North polarity. Shading ring polarity is zero, so the stator pole moves toward the shading ring.

Position 3 Top of stator: Main pole polarity is less strong North pole. Shaded pole is stronger North polarity. Stator pole moves more toward the shaded pole.

Position 4 Top of stator: Main pole has no current and no flux. Shaded pole has stronger North polarity, so stator pole moves toward the shaded pole. This produces a counterclockwise rotation of north pole flux.

To reverse the direction of rotation on shaded pole motors, the relationship between the shading ring and the pole face must be reversed, as the field always moves from the unshaded to the shaded portion of the stator pole. Another set of shading rings on the opposite side may be shorted. A reversible shaded pole motor has two sets of shading rings. When one set is shorted, the other set is open.

Controlling the speed of shaded pole motors is easy. Either alter the voltage that is applied to the stator winding to produce less voltage per turn on the winding, or change the number of turns by using tapped windings and maintaining the same applied voltage. Either method changes the voltage per turn on the motor. Less voltage per turn means less flux, more slip, and a slower speed under load.

Shaded pole motors are typically used for fans where the blades are directly mounted to the rotor shaft and the air passes over the motor. These fans require little starting torque, and the air over the motor helps keep it cool. If these motors run hot or are in a high vibration area, the shading rings (which are soldered rings of conductors) can open up and cause the motor to fail.

SHADING RINGS

FIGURE 19–24 Photo of one style of shaded pole motor.

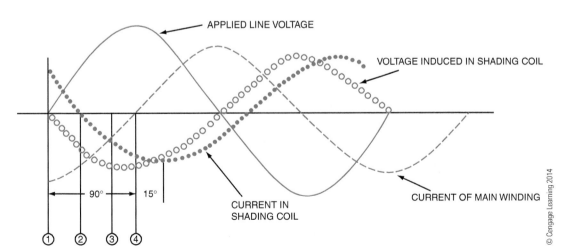

FIGURE 19–25 Shaded pole motor voltage and current waveforms: line voltage, main coil current, and induced shading coil waveforms.

SUMMARY

The single-phase induction motor is one of the most used residential and light commercial motors. Each application dictates the correct motor style to use. All these motors use the concept of taking one phase, or one sinewave, and shifting the effects of the currents through the coils to create a moving magnetic field. The split-phase and capacitor-start motors use a starting switch to disconnect the starting windings from the line after the motor is up to running speed. Two-capacitor motors use multiple capacitors or variations of two-value capacitors to create a starting and a running circuit. All *NEC*® regulations that apply to three-phase motors apply to single-phase motors. Many exceptions apply only to small-horsepower motors.

ACHIEVEMENT REVIEW

1. List the essential parts of a split-phase induction motor.

   _____

   _____

   _____

   _____

2. What happens if the centrifugal switch contacts fail to reclose when the motor
   stops? _____

   _____

   _____

   _____

   _____

3. Explain how the direction of rotation of a split-phase induction motor is reversed.

   _____

   _____

   _____

   _____

   _____

   _____

4. A split-phase induction motor has a dual-voltage rating of 115/230 volts. The motor has two running windings, each of which is rated at 115 volts, and one starting winding rated at 115 volts. Draw a schematic diagram of this split-phase induction motor connected for a 230-volt operation.

5. Draw a schematic connection diagram of the split-phase induction motor in question 4, connected for a 115-volt operation.

6. A split-phase induction motor has a dual-voltage rating of 115/230 volts. The motor has two running windings, each of which is rated at 115 volts, and one starting winding rated at 115 volts. Draw a schematic diagram of this split-phase induction motor connected for a 230-volt operation.

7. What is the primary difference between a split-phase induction motor and a capacitor-start, induction-run motor?_____

_____

_____

_____

_____

8. If the centrifugal switch fails to open as a split-phase motor accelerates to its rated speed, what happens to the starting winding?_____

_____

_____

_____

_____

9. Describe one limitation of a capacitor-start, induction-run motor.

_____

_____

_____

10. Insert the correct word or phrase to complete each of the following statements.
    a. The capacitor used with a capacitor-start, induction-run motor is used only to improve
       the _____.
    b. A capacitor-start, induction-run motor has a better starting torque than the

       _____.

11. Shaded pole motors are _____
    a. high-torque motors.
    b. low percent, speed-regulation motors.
    c. used for large horsepower loads.
    d. used for low-torque, low-horsepower loads.

UNIT 20

SPECIAL MOTORS AND APPLICATIONS

OBJECTIVES

After studying this unit, the student should be able to

- describe the operations of stepper motors.

- determine two types of Servo motors.

- explain how motion control and feedback are obtained with servo motors.

- explain how braking and regeneration works.

- describe the operation of a simple selsyn system and a differential selsyn system.

- list several advantages of a selsyn system.

STEPPER MOTORS

Stepper motors are specialized motors that also create incremented steps of motion rather than a smooth unbroken rotation. The basic stepper concept is explained using a permanent magnet on the rotor with two sets of poles, as shown in Figure 20–1. As the stator is energized with pulses of DC, the permanent magnet rotor is repelled or attracted to line up with the stator magnetic poles. A stepper controller provides the pulses, as illustrated in Figure 20–2.

The electronic controller provides timing and sequencing of the motor, but it operates electronically to provide circuit closure, as shown in Figure 20–1. For example, by moving switches 1 and 2 to position A or B, the top poles or the side poles can be reversed. By following the first switch sequence where both switches are set to A, the top and right poles become north magnetic polarity and the rotor aligns between the poles. Step two changes switch 1 to B. If power is left on the top poles, the rotor aligns top to bottom. When switch 1 connects to point B, the rotor again moves, the south rotor pole aligning between the top and left poles.

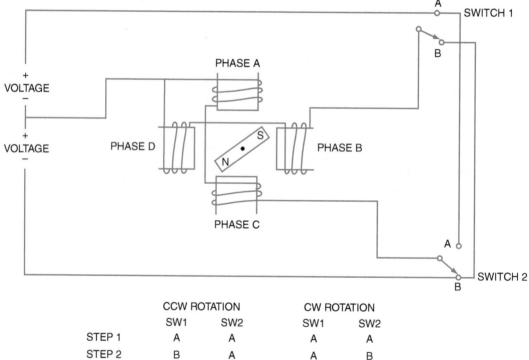

| | CCW ROTATION | | CW ROTATION | |
	SW1	SW2	SW1	SW2
STEP 1	A	A	A	A
STEP 2	B	A	A	B
STEP 3	B	B	B	B
STEP 4	A	B	B	A
STEP 5	A	A	A	A

© Cengage Learning 2014

FIGURE 20–1 Stepper motor showing switch position and DC voltage source.

FIGURE 20-2 Stepper motor and associated electronic controller.

This results in counterclockwise (CCW) rotation. To reverse the direction of rotation, use the second set of steps. Note that by changing the sequence and length of time the coils are energized, and the direction and speed of the steps are controlled.

The rotor could be one of three different styles: the variable-reluctance rotor, the permanent-magnet (PM) rotor, or a combination of the two—a hybrid rotor. Rather than being simply two magnetic poles, the rotor is many magnetic poles lined up with the teeth on the rotor. The rotor's teeth are spaced so that only one set of teeth remains in perfect alignment with the stator poles at any one time.

By taking the number of times the power must be applied to the stator poles to move one tooth through 360° of mechanical rotation, you can calculate the step angle. For example, if the stator needs 200 pulses of power to move one tooth 360°, then divide 360 by 200 to get 1.8° of motion per step. The motor will move 1.8° per pulse of stator power. Steppers are available step angles of 90°, 45°, 15°, 7.5°, 1.8°, and 0.9°. The resolution of the motor is a measure of how fine the steps are divided. Resolution is determined by dividing 360° by the step angle. The 1.8° step-angle motor takes 200 steps to move around a full revolution, so the resolution is 200. Step angle and resolution are inversely proportional to each other.

The PM rotor was used for the example on step motion control. These PM rotors are used with the four pole pieces to provide either 90° or 45° of step angle. This allows the motors to turn at higher speed, but with less resolution. The poles can be physically larger because there are only four of them; thus, the stator windings can carry higher current. Higher current capability means that more torque is available.

SERVO MOTORS

Many manufacturing systems and robotic material handling systems use Servomotors to position controls for fast precision movement. Servo motors are small PM motors that use lightweight armatures to provide quick start and stop operations without a lot of inertia to overcome. See Figure 20–3 for a sample of a PM motor with four poles.

FIGURE 20–3 Servo motor used to produce robot joint movement.

At slow speeds of rotation, the standard PM servomotors have a tendency to create cogging; that is, they tend to ratchet through the rotation rather than have smooth rotation. This is a result of the iron in the armature being magnetized and reacting with the PM on the stator pole pieces.

Servo Motor and Controllers

The two general types of AC servo-motors are considered either synchronous or asynchronous. Synchronous Servos do not use induction type rotors as in the most common types of AC motors. They do not rely on an AC induction rotor that has current induced into it. Instead they rely on a rotor that is made of PMs. The concept is to make the rotor as light as possible so that the rotor itself has little mass and therefore little inertia to overcome as it starts and stops.

As the stator field is energized, the rotor follows the rotating magnetic field of the stator at the same (synchronous) speed. As the stator field stops, the rotor stops too. With a PM rotor, less heat is produced in the motor because it has no rotor current. Also because no rotor current is needed from the line, the motors are more efficient. To control where the rotor is with respect to the stator, feedback is required to the motor controller. An encoder is typically used and the information is fed back to the controller within one revolution on the synchronous Servo.

Asynchronous Servos rely on induction to supply current and therefore magnetic flux to the rotor. The rotor is like a typical squirrel cage, with efforts to keep it as light as possible to keep inertia low. As in regular induction motors, the rotating field of the stator cuts the rotor conductors and induces current into the rotor, which creates flux in the rotor. This magnetic field is then pushed and pulled around the stator. As with all induction motors, there is slip, as the rotor's poles must fall behind the stator poles. Again, the controller must know the actual position of the rotor for precise speed and position controls.

Servo motors are used for exact speed, torque, and positioning of driven machines. Because speed is the major parameter, the speed of the rotating magnetic stator field is controlled. In most cases, this is done by controlling the frequency of the applied AC to the stator. To do this, an electronic controller is used to create variable AC frequency and voltage. A pulse-width modulation (PWM) system is used to control the desired frequency and the corresponding voltage needed to drive the motor. PWMs are described in Unit 15. Be sure to check that a particular Servo motor is designed to fit a particular controller and that the motor is designed to work with PWM drives. The torque, or twisting effort of the motor, is also a major parameter. The rated torque is often listed at full speed, at half speed, and at zero speed (stalled). The stall torque is the capability to hold a load in place when there is no rotational speed. This stall point has an associated stall current and often lists the associated rise in temperature as the needed current is applied to hold the load stationary. The winding insulation temperature on the motor is designed to withstand the hot spot temperature of the winding as it holds indefinitely the stall current and the associated torque.

Duty cycles are used as a parameter as to how long and how hard the motor can run without cooling down. The duty cycles are arranged from S1 to S9. S1 is a continuous rating, meaning no rest or "off" time is required. An S3 can run for 4 out of every 10 minutes. An S6 runs 4 minutes at full load and 6 minutes at stalled load. Feedback of speed and position of the rotor is accomplished by use of encoders and resolvers. An incremental encoder is used for basic information (see Figure 20–4). The encode signal tells the controller whether the rotor is accelerating, at set speed, or decelerating (see Figure 20–5). By counting the number of pulses, the controller can determine how many revolutions the rotor has made since the start command was issued. This system closes the loop between the desired command sent to the motor and the actual rotor response. A variation on the encoder design and reading created a reference pulse for the controller to know where asynchronous Servos rotors are in their real revolutions.

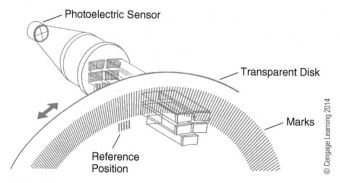

FIGURE 20-4 One style of encoder uses photoelectric sensors and incremental marks to track motion.

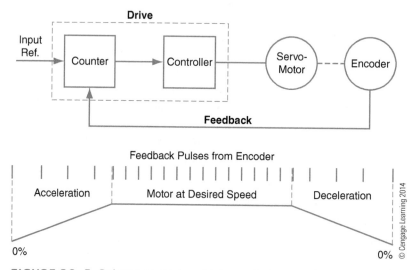

FIGURE 20-5 Pulses are read and interpreted as acceleration, deceleration, or constant speed.

There is less precision because there has to be one full revolution for the encoder to register the reference mark. As the need for precision increases, more sophisticated techniques are used to keep track of the exact position of the rotor. The controller uses a more complex system for following the rotor movement and keeping track of the total number of revolutions from an initial reference point.

Resolver methods to track the rotor movement can also be used. The resolver uses the technique of a rotating magnet that induces a sine wave into the receiver. The rotor uses two magnetic fields spaced 90° apart. As they pass the receiver, they each induce a sine wave. By comparing the two waveforms, the controller calculates the speed, direction, and position of the rotor (see Figure 20–6).

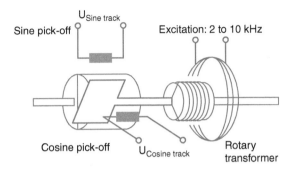

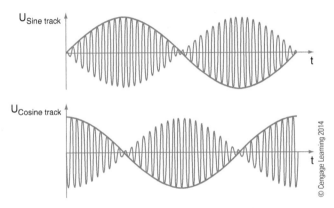

FIGURE 20–6 A resolver uses a magnetic field to track motion.

Braking and Regeneration

Motors have four general conditions: start and accelerate, run at set speed, decelerating to stop, and stopped or stalled. Decelerating or slowing down can be gradual with removal of power and coasting to a stop, or the deceleration can be controlled and finally a brake applied to hold the rotor position. The process of slowing the motor with a mechanical load attached has the effect of the load inertia pulling the rotor faster than the rotating magnetic field provides. This overhauling effect is known as *negative torque*. As the rotor spins past the magnetic field of the stator, the motor becomes a generator, or produces a regenerated voltage. This voltage is actually generated back into the stator as a reverse voltage from the applied voltage. This regeneration can be used to slow the motor, just as a generator would slow down as it does more electrical work. If the electrical energy is fed back to the controller, the voltage could cause damage to the DC converter section of the controller. This is the front end of the controller that converts AC to DC for use by the PWM inverter section. By installing a braking resistor across the converter DC bus, the excess energy can be dissipated by closing the circuit to the resistor, in this case through an insulated gate bipolar transistor (IGBT), as shown in Figure 20–7.

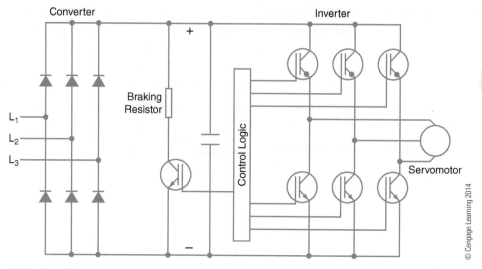

FIGURE 20–7 A braking resistor can be used to dissipate energy during braking.

Another system allows the regenerated power to be returned to the power source. This method uses an SCR rectifier bridge to rectify the AC supply for the DC bus. A separate SCR rectifier bridge is mounted in reverse to the first and is turned on when there is regeneration. The regenerated power is now returned directly to the power source (see Figure 20–8).

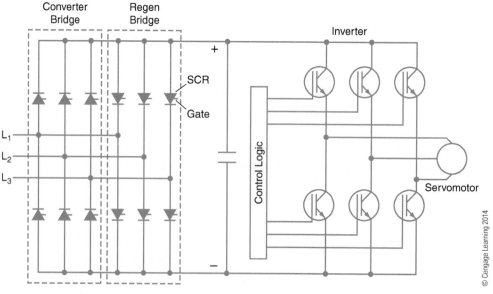

FIGURE 20–8 Regenerated power is returned to power source.

Servo Disc motor is a motor with a different design. The rotor is a nonconductive disc that has layers of copper "printed" on it, which leads to another name for the same motor: printed circuit motor. (See Figure 20–9.) The PMs that form the stator are mounted on either side of the disc rotor. The conductors on the rotor are brought out to one side of the disc, and external connections are made through small brushes that ride on the disc. Because the conductors are under a PM pole piece, the current in the rotor creates a magnetic flux that reacts with the pole piece to create torque. The number of pole pieces and the corresponding conductors multiplies this twisting torque. The turning effort is smooth at all speeds because there is no iron in the armature to create the cogging effect as seen in traditional PM motors. Also, because of the thin, not steel rotor, the Servo Disc motor has very fast acceleration. It can accelerate to 3000 RPM in about 60° of rotation, or 1/6 of one revolution. The same feature of the rotor allows it to stop and reverse very quickly as well. As with other DC motors, the current through the rotor controls the speed. The rotor resistance is, of course, a fixed value, so the voltage control to the armature is varied to control speed, and voltage is reversed to provide reverse direction.

As with the AC electronic controls, PWM is used to control both the voltage level and the frequency. The frequency is maintained near 20 kH of pulsating DC. By controlling the "on" time of the pulse, the voltage is varied, and therefore the speed is controlled. By controlling the direction of the DC polarity, the direction is controlled. See Figure 20–10 for a photo of a Servo Disc motor.

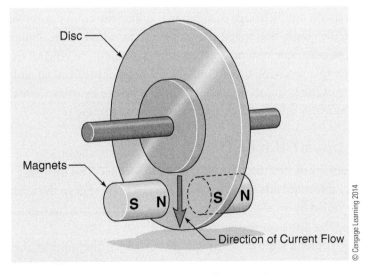

FIGURE 20-9 Basic construction of a ServoDisc motor.

FIGURE 20–10 The DC ServoDisc motor can be constructed with small, thin cases.

SELSYN MOTORS

The word *selsyn* is an abbreviation of the term *self-synchronous*. Selsyn units are special AC motors used primarily in applications requiring remote control. Small selsyn units transmit meter readings or values of various types of electrical and physical quantities to distant points. For example, the captain on the bridge of a ship may adjust the course and speed of the ship; at the same moment, the course and speed changes are transmitted to the engine room by selsyn units. On the engine telegraph system, mechanical positioning of a control transmits electrical angular information to a receiving unit. Similarly, readings of mechanical and electrical conditions in other parts of the ship can be recorded on the bridge by selsyn units. These units are also referred to as *synchros,* and are known by various trade names.

STANDARD SELSYN SYSTEM

A selsyn system consists of two three-phase induction motors. The normally stationary rotors of these induction motors are interconnected so that a manual shift in the rotor position of one machine is accompanied by an electrical rotor shift in the other machine in the same direction and of the same angular displacement as the first unit.

Figure 20–11 shows a simple selsyn system for which the units at the transmitter and receiver are identical. The rotors of these units are two pole and must be excited from the same AC source. Three leads between the transmitter and the receiver units connect the three-phase

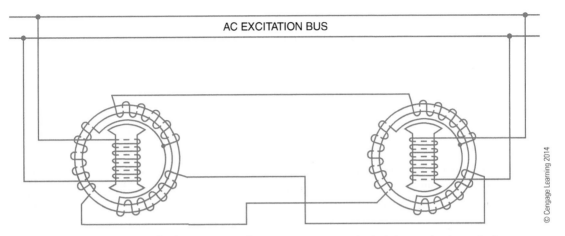

FIGURE 20–11 Diagram of selsyn motors showing interconnected stator and rotor windings connected to excitation source.

FIGURE 20–12 Wound rotor with oscillation damper and slip rings for selsyn units.

stator windings to each other. The rotor of each machine is called the *primary,* and the three-phase stator winding of each machine is called the *secondary*. A rotor for a typical selsyn unit is shown in Figure 20–12.

When the primary excitation circuit is closed, an AC voltage is impressed on the transmitter and receiver primaries. If both rotors are in the same position with respect to their stators, no movement occurs. If the rotors are not in the same relative position, the freely movable receiver rotor will turn to assume the same position as the transmitter rotor. If the transmitter rotor is turned, either manually or mechanically, the receiver rotor will follow at the same speed and in the same direction.

The self-synchronous alignment of the rotors is the result of voltages induced in the secondary windings. Both rotors induce voltages into the three windings of their stators. These voltages vary with the position of the rotors. If the two rotors are in the same relative position, the voltages induced in the transmitter and receiver secondaries are equal and opposite. In this condition, current does not exist in any part of the secondary circuit.

If the transmitter rotor is moved to another position, the induced voltages of the secondaries are no longer equal and opposite, and currents are present in the windings. These currents establish a torque that tends to return the receiver rotor to a synchronous position. Because the receiver rotor is free to move, it makes the adjustment. Any movement of the transmitter rotor is accompanied immediately by an identical movement of the receiver rotor.

DIFFERENTIAL SELSYN SYSTEM

Figure 20–13 shows a diagram of the connections of a differential selsyn system consisting of a transmitter, a receiver, and a differential unit. This system produces an angular indication of the receiver. The indication is either the sum or difference of the angles existing at the transmitter and differential selsyns. If two selsyn generators, connected through a differential selsyn, are moved manually to different angles, the differential selsyn indicates the sum or difference of their angles.

A differential selsyn has a primary winding with three terminals. Otherwise, it closely resembles a standard selsyn unit. The three primary leads of the differential selsyn are brought out to collector rings. The unit has the appearance of a miniature wound-rotor, three-phase induction motor. The unit, however, normally operates as a single-phase transformer.

The voltage distribution in the primary winding of the differential selsyn is the same as that in the secondary winding of the selsyn exciter. If any one of the units is fixed in position and a second unit is displaced by a given angle, then the third unit (which is free to rotate) turns through the same angle. The direction of rotation can be reversed by interchanging any pair of leads on either the rotor or stator winding of the differential selsyn.

If any two of the selsyns are rotated simultaneously, the third selsyn turns through an angle equal to the algebraic sum of the movements of the two selsyns. The algebraic sign of this value depends on the rotation direction of the rotors of the two selsyns, as well as the phase rotation of their windings.

The excitation current of the differential selsyn is supplied through connections to one or both of the standard selsyns to which the differential selsyn is connected. In general, the excitation current is supplied to the primary winding only. In this case, the selsyn connected to the differential stator supplies this current and must be able to carry the extra load without overheating. A particular type of selsyn, known as an *exciter selsyn*, is used to supply the current. The exciter selsyn can function in the system either as a transmitter or a receiver.

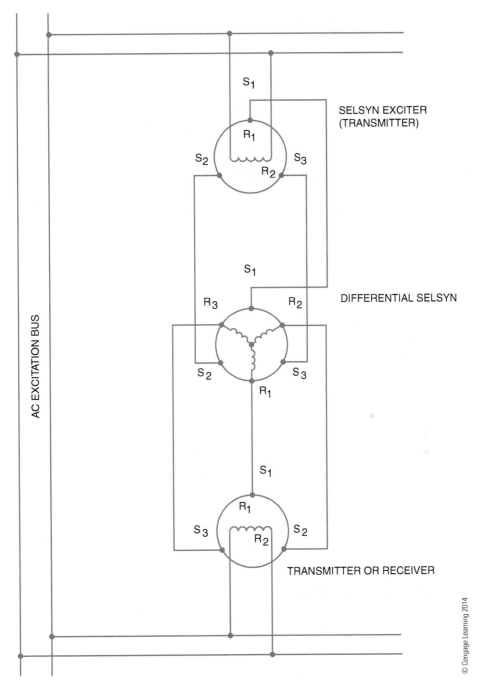

FIGURE 20–13 A schematic diagram of differential selsyn connections.

ADVANTAGES OF SELSYN UNITS

Selsyn units are compact and rugged, and provide accurate and very reliable readings. Because of the comparatively high torque of the selsyn unit, the indicating pointer does not oscillate as it swings into position. Internal mechanical dampers are used in selsyn receivers to prevent oscillation during the synchronizing procedure and to reduce any tendency of the receiver to operate as a rotor. The operation of the receiver is smooth and continuous and is in agreement with the transmitter. In addition, the response of the receiver to changes in position at the transmitter is very rapid.

In the event of a power failure, the indicator of the receiver resets automatically with the transmitter when power is received. Calibration and time-consuming checks are unnecessary. A number of advantages are offered by selsyn systems:

- The indicators are small and compact and can be located where needed.
- The simple installation requires running a few wires and bolting the selsyn units in place.
- Selsyn units can be used to indicate either angular or linear movement.
- Selsyn units control the motion of a device at a distant point by controlling its actuating mechanism.
- One transmitter may be used to operate several receivers simultaneously at several distant points.

SUMMARY

Stepper motors are used for incremental motion control. They produce motion in steps as a magnetic field moves as determined by a stepper controller. Servo motors also provide precision motion control and quick start and stop operation as needed in positioning controls. Two styles of servos are synchronous and asynchronous motors. Feed back as to position, speed and rotational direction can be acquired through different techniques. Braking and holding of motors is essential to positioning motors to move and hold loads as directed by the controller. The selsyn system is also referred to as a synchro system. The self-synchronous system allows one rotor to act as a transmitter and the other to act as a receiver to either directly follow the sender, or follow at a prescribed offset angle.

ACHIEVEMENT REVIEW

Select the correct answer for each of the following statements, and place the corresponding letter in the space provided.

1. Selsyn transmitters and receivers resemble _____
   a. repulsion-induction motors.
   b. three-phase, two-pole induction motors.
   c. three-phase, four-pole induction motors.
   d. synchronous machines.

2. When the primary excitation circuit is closed, AC voltage is impressed on the

   _____
   a. transmitter and receiver primaries.
   b. transmitter rotor and the transmitter stator windings.
   c. transmitter rotor and the receiver stator windings.
   d. stator windings of both instruments.

3. A differential selsyn unit differs from a selsyn transmitter or receiver in that it requires

   _____
   a. three-phase power for excitation.
   b. an AC line connection to the stator winding.
   c. DC on the rotor winding.
   d. three connections to the rotor winding.

4. If the rotors of the two selsyn units in a selsyn-indicating system are in exactly corresponding positions, the current in the secondary winding is _____
   a. within quadrature with the primary current.
   b. in phase with the primary current.
   c. zero.
   d. less than the normal primary current.

5. Selsyn units are also referred to as _____
   a. synchros.
   b. induction motors.
   c. wound-rotor motors.
   d. all of the above

6. The stator of the transmitter is directly connected to the stator of the receiver unit when a differential is not used. _____
   a. true
   b. false

7. In the transmitter and receiver system, the AC excitation is applied to the _____
   a. stator winding.
   b. stator and the rotor windings.
   c. rotor winding only.
   d. none of the above

8. Cite several advantages of a selsyn system._____

    _____

    _____

    _____

    _____

9. The purpose of a stepper motor is to
    a. provide incremental steps of motion.
    b. provide steps of rotation that are imperceptible.
    c. create a vertical motion rather than a rotational motion.
    d. create a stair-step output from the controller.

10. A synchronous Servos uses
    a. wound rotors for rotor construction.
    b. PM rotors.
    c. shaded pole rotors.
    d. squirrel-cage rotors.

11. Asynchronous Servos use
    a. wound rotors for rotor construction.
    b. PM rotors.
    c. shaded pole rotors.
    d. squirrel-cage rotors.

12. When a Servo motor is slowing down, the load inertia sometimes causes the motor to spin faster than desired. This is called
    a. negative torque.
    b. over-speed torque.
    c. regenerative torque.
    d. dynamic braking inertia.

UNIT 21

AC SERIES AND REPULSION MOTORS

After studying this unit, the student should be able to

- describe the basic operation of a universal motor.
- explain how a single-field compensated universal motor operates.
- explain how a two-field compensated universal motor operates.
- describe two ways in which universal motors are compensated for excessive armature reaction under load.
- state the reasons why DC motors fail to operate satisfactorily from an AC source.
- describe the basic steps in the operation of the following types of motors:
- repulsion motor.
- repulsion-start, induction-run motor.
- repulsion-induction motor.
- state the basic construction differences among the motors in the preceding list.
- compare the motors in the preceding list with regard to starting torque and speed performance.

The electrician may consider a typical DC series motor or a DC shunt motor for operation on AC power supplies. It appears that such operation is possible because reversing the line terminals to a DC motor reverses the current and magnetic flux in both the field and armature circuits. As a result, the net torque of the motor operating from an AC source is in the same direction.

However, the operation of a DC shunt motor from an AC source is impractical because the high inductance of the shunt field causes the field current and the field flux to lag the line voltage by almost 90°. The resulting torque is very low.

A DC series motor also fails to operate satisfactorily from an AC source because of the excessive heat developed by eddy currents in the field poles. In addition, an excessive voltage drop occurs across the series field windings due to high reactance.

To reduce the eddy currents, the field poles can be laminated. To reduce the voltage loss across the field poles to a minimum, a small number of field turns can be used on a low reactance core operated at low flux density. A motor with these revisions operates on either AC or DC and is known as a universal motor. Universal motors in small fractional horsepower sizes are used in household appliances and portable power tools.

Repulsion motors are divided into three distinct classifications: the repulsion motor; the repulsion-start, induction-run motor; and the repulsion-induction motor. Although these motors are similar in name, they differ in construction, operating characteristics, and industrial applications.

CONCENTRATED-FIELD UNIVERSAL MOTORS

A concentrated-field universal motor is usually a salient-pole motor with two poles and a winding of relatively few turns. The poles and winding are connected to give opposite magnetic polarity. A field yoke of this type of motor is shown in Figure 21–1.

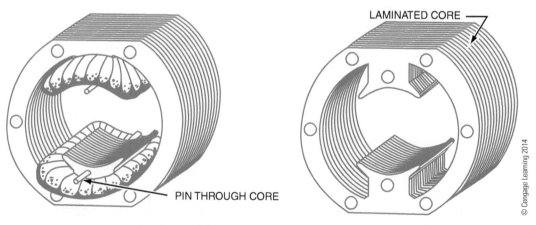

FIGURE 21–1 Field core of a two-pole universal motor.

DISTRIBUTED-FIELD UNIVERSAL MOTORS

The two types of distributed-field universal motors are the single-field compensated motor and the two-field compensated motor. The field windings of a two-pole, single-field compensated motor resemble the stator winding of a two-pole, split-phase AC motor. A two-field compensated motor has a stator containing a main winding and a compensating winding spaced 90 electrical degrees apart. The compensating winding reduces the reactance voltage developed in the armature by the alternating flux when the motor operates from an AC source. Figure 21–2 is the schematic diagram of a compensated universal motor.

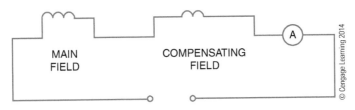

FIGURE 21–2 Schematic diagram of a compensated universal motor.

THE ARMATURE

The armature of a typical universal motor resembles the armature of a typical DC motor except that a universal motor armature is slightly larger for the same horsepower output.

CONSTRUCTION FEATURES OF UNIVERSAL MOTORS

The frames of universal motors are made of aluminum, cast iron, or rolled steel. The field poles are generally bolted to the frame. Field cores consist of laminations pressed together and held by bolts. The armature core is also laminated and has a typical commutator and brushes. End plates resemble those of other motors except that in many universal motors, one end plate is cast as part of the frame. Both ball and sleeve bearings are used in universal motors.

SPEED CONTROL

Universal motors operate at approximately the same speed on DC or single-phase AC. Because these motors are series wound, they operate at excessive speed at a no-load condition. As a result, they are usually permanently connected by gears to the devices being driven. Universal motors are speed-regulated by inserting resistance in series with the motor. The resistance may be tapped resistors, rheostats, or tapped nichrome wire coils wound over a single field pole. In addition, speed may be controlled by varying the inductance through taps on one of the field poles. Gear boxes are also used.

Speed control of series motors can also be accomplished by using electronic speed controls. The concept is the same as used in series voltage drops; that is, the voltage to the motor is reduced to give a reduced speed. This can be done by using SCRs or triacs to alter the voltage available to the motor.

DIRECTION OF ROTATION

The direction of rotation of any series-wound motor can be reversed by changing the direction of the current in either the field or the armature circuit. Universal motors are sensitive to brush position, and severe arcing at the brushes results from changing the direction of rotation without shifting the brushes to the neutral (sparkless) plane or adding a compensating winding.

CONDUCTIVE COMPENSATION

AC motors rated at more than 0.5 hp are used to drive loads requiring a high starting torque. Two methods are used to compensate for excessive armature reaction under load. In the conductively compensated type of motor, an additional compensating winding is placed in slots cut directly into the pole faces. The strength of this field increases with an increase in load current and thus minimizes the distortion of the main field flux by the armature flux (called *armature reaction*, discussed in Unit 1). The compensating winding is connected in series with the series field winding and the armature, as shown in Figure 21–3. Although conductively compensated motors have a high starting torque, the speed regulation is poor. A wide range of speed control is possible with the use of resistor-type starter controllers.

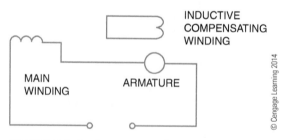

FIGURE 21–3 Connections for an inductively compensated universal motor.

INDUCTIVE COMPENSATION

Armature reaction in AC series motors may also be compensated with an inductively coupled winding, which acts as a short-circuited secondary winding of a transformer. This winding is placed so that it links the cross-magnetizing flux of the armature, which acts as the primary winding of a transformer. Figure 21–3 shows the schematic diagram of an inductively compensated universal motor. Because the magnetomotive force of the secondary is nearly opposite in phase and equal in magnitude to the primary magnetomotive force, the compensating winding flux nearly neutralizes the armature cross flux. This type of motor cannot be used on DC. Because of its dependency on induction, the operating characteristics of an inductively compensated motor are very similar to those of the conductively compensated motor.

REPULSION MOTOR

A repulsion motor basically consists of the following parts:

- *Laminated stator core with one winding.* This winding is similar to the main or running winding of a split-phase motor. The stator is usually wound with four, six, or eight poles.
- *Rotor consisting of a slotted core into which a winding is placed.* The rotor is similar in construction to the armature of a DC motor. Thus, the rotor is called an armature. The coils that make up this armature winding are connected to a commutator. The commutator has segments or bars parallel to the armature shaft.
- *Carbon brushes contacting with the commutator surface.* The brushes are held in place by a brush-holder assembly mounted on one of the end shields. The brushes are connected by heavy copper jumpers. The brush-holder assembly may be moved so that the brushes can make contact with the commutator surface at different points to obtain the correct rotation and maximum torque output. Brush arrangement types include *brush riding,* where the brushes are in contact with the commutator surface at all times, and *brush lifting,* in which the brushes lift at approximately 75% of the rotor speed.
- *Two cast steel end shields.* These shields house the motor bearings and are secured to the motor frame.
- *Two bearings supporting the armature shaft.* The bearings center the armature with respect to the stator core and windings. The bearings may be sleeve bearings or ball-bearing units.
- *Cast steel frame.* The stator core is pressed into this frame.

Operation of a Repulsion Motor

The connection of the stator winding in a repulsion motor to a single-phase line causes a field to be developed by the current in the stator windings. This stator field induces a voltage and a resultant current in the rotor windings. If the brushes are placed in the proper position on the commutator segments, the current in the armature windings sets up proper magnetic poles in the armature.

These armature field poles have a set relationship to the stator field poles. That is, the magnetic poles developed in the armature are offset from the field poles of the stator winding by about 15 electrical degrees. Furthermore, because the instantaneous polarity of the rotor poles is the same as that of the adjacent stator poles, the repulsion torque created causes the rotation of the motor armature.

The three diagrams of Figure 21–4 show the importance of the brushes being in the proper position to develop maximum torque. In Figure 21–4(A), no torque is developed when the brushes are placed at right angles to the stator poles. This is because the equal induced voltages in the two halves of the armature winding oppose each other at the connection between the two sets of brushes. Because no current is in the windings, flux is not developed by the armature windings. This position is called *soft neutral*.

In Figure 21–4(B), the brushes are in a position directly under the center of the stator poles. A heavy current exists in the armature windings with the brushes in this position, but there is still no torque. The heavy current in the armature windings sets up poles in the armature. However, these poles are centered with the stator poles, and a torque is not created in either a clockwise or counterclockwise direction. This position is called *hard neutral*.

In Figure 21–4(C), the brushes have shifted from the center of the stator poles 15 electrical degrees in a counterclockwise direction. Thus, magnetic poles of like polarity are set up in the armature. These poles are 15 electrical degrees in a counterclockwise direction from the stator pole centers. A repulsion torque is created between the stator and the rotor field poles of like polarity. The torque causes the armature to rotate in a counterclockwise direction. A repulsion machine has a high starting torque with a small starting current, and a rapidly decreasing speed with an increasing load.

The direction of rotation of a repulsion motor is reversed if the brushes are shifted 15 electrical degrees from the stator field pole centers in a clockwise direction, shown in Figure 21–5. As a result, magnetic poles of like polarity are set up in the armature. These poles are 15 electrical degrees in a clockwise direction from the stator pole centers. Repulsion motors are used principally for constant-torque applications, such as printing-press drives, fans, and blowers.

REPULSION-START, INDUCTION-RUN MOTOR

A second type of repulsion motor is the repulsion-start, induction-run motor. In this type of motor, the brushes contact the commutator at all times. The commutator of this motor is the more conventional axial form.

A repulsion-start, induction-run motor consists basically of the following parts:

- *Laminated stator core.* This core has one winding similar to the main or running winding of a split-phase motor.

- *Rotor consisting of a slotted core into which a winding is placed.* The coils that make up the winding are connected to a commutator. The rotor core and winding are similar to the armature of a DC motor. Thus, the rotor is called an armature.

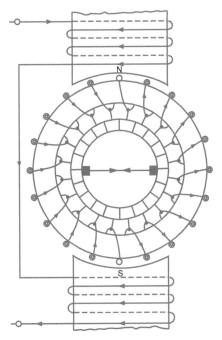

A. NO TORQUE CREATED, EQUAL VOLTAGE VALUES
    OPPOSE EACH OTHER (SOFT NEUTRAL)

B. NO TORQUE EVEN THOUGH CURRENT VALUE
    IN ARMATURE IS HIGH (HARD NEUTRAL)

C. COUNTERCLOCKWISE ROTATION;
    BRUSHES IN CORRECT POSITION

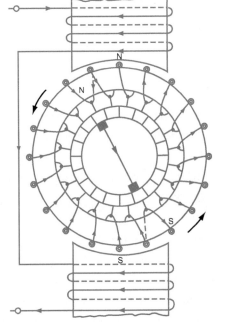

FIGURE 21–4 Repulsion motor operation.

© Cengage Learning 2014

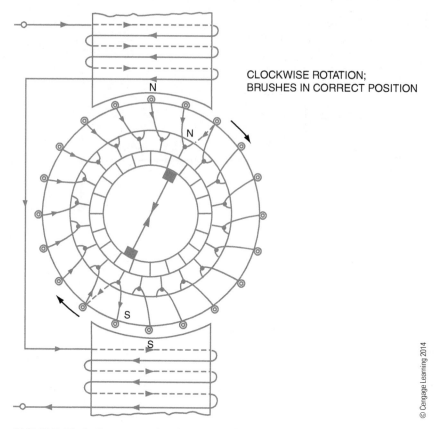

CLOCKWISE ROTATION;
BRUSHES IN CORRECT POSITION

FIGURE 21–5 Reversing the direction of rotation of a repulsion motor.

- *Centrifugal device.*
  a. In the brush-lifting type of motor, a centrifugal device lifts the brushes from the commutator surface at 75% of the rated speed. This device consists of governor weights, a short-circuiting necklace, a spring barrel, spring, push rods, brush holders, and brushes. Although high in initial cost, this device does save wear and tear on brushes, and runs quietly. Figure 21–6 is an exploded view of the armature, radial commutator, and centrifugal device of the brush-lifting type of repulsion-start, induction-run motor.
  b. The brush-riding type of motor also contains a centrifugal device that operates at 75% of the rated speed. This device consists of governor weights, a short-circuiting necklace, and a spring barrel. The commutator segments are short-circuited by this device, but the brushes and brush holders are not lifted from the commutator surface.
- *Commutator.* The brush-lifting type of motor has a radial type of commutator (Figure 21–6). The brush-riding type of motor has an axial commutator (Figure 21–7).

© Cengage Learning 2014

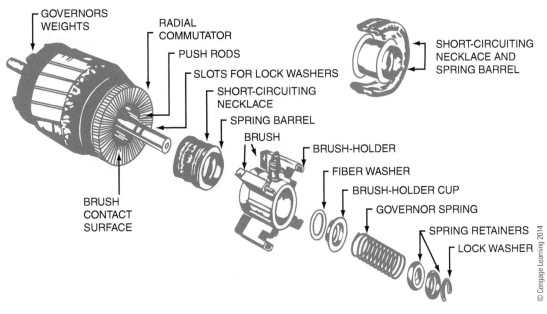

FIGURE 21–6 An exploded view of a radial commutator and centrifugal brush-lifting device for a repulsion-start, induction-run motor.

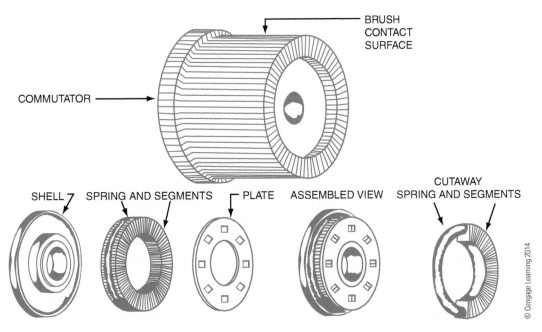

FIGURE 21–7 An exploded view of a short-circuiting device for a brush-riding, repulsion-start, induction-run motor.

- *Brush-holder assembly.*

  a. The brush-holder assembly for the brush-lifting type of motor is arranged so that the centrifugal device can lift the brush holders and brushes clear of the commutator surface.

  b. The brush-holder assembly for the brush-riding type of motor is the same as that of a repulsion motor.

- *End shields, bearings, and motor frame.* The parts are the same as those of a repulsion motor.

Operation of the Centrifugal Mechanism

Refer to Figure 21–7 to identify the components of the centrifugal mechanism. The operation of this device consists of the following steps. As the push rods of the centrifugal device move forward, they push the spring barrel forward. This allows the short-circuiting necklace to make contact with the radial commutator bars; thus, all are short-circuited. At the same time, the brush holders and brushes are moved from the commutator surface. As a result, there is no unnecessary wear on the brushes and the commutator surface, and there are no objectionable noises caused by the brushes riding on the radial commutator surface.

The short-circuiting action of the governor mechanism and the commutator segments converts the armature to a form of squirrel-cage rotor, and the motor operates as a single-phase induction motor. In other words, the motor starts as a repulsion motor and runs as an induction motor.

In the brush-riding type of motor, an axial commutator is used. The centrifugal mechanism (Figure 21–7) consists of a number of copper segments that are held in place by a spring. This device is placed next to the commutator. When the rotor reaches 75% of the rated speed, the centrifugal device short-circuits the commutator segments. The motor then continues to operate as an induction motor.

Operation of a Repulsion-Start, Induction-Run Motor

The starting torque is good for either the brush-lifting type or the brush-riding type of repulsion-start, induction-run motor. Furthermore, the speed performance of both types of motors is very good because they operate as single-phase induction motors.

Because of the excellent starting and running characteristics for both types of repulsion-start, induction-run motors, they have been used for a variety of industrial applications, including commercial refrigerators, compressors, and pumps.

The direction of rotation for a repulsion-start, induction-run motor is changed in the same manner as that for a repulsion motor, that is, by shifting the brushes past the stator pole center 15 electrical degrees. The symbol in Figure 21–8 represents both a repulsion-start, induction-run motor and a repulsion motor.

Many repulsion-start, induction-run motors are designed to operate on 115 volts or 230 volts. These dual-voltage motors contain two stator windings. For 115-volt operation, the stator windings are connected in parallel; for 230-volt operation, the stator windings are connected in series. The diagram in Figure 21–9 represents a dual-voltage, repulsion-start, induction-run motor. The connection table shows how the leads of the motor are connected for either 115-volt operation or 230-volt operation. These connections also can be used for dual-voltage repulsion motors.

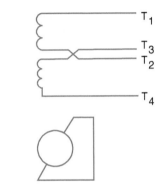

FIGURE 21-8 Schematic diagram symbol of a repulsion-start,induction-run motor and a repulsion motor.

© Cengage Learning 2014

	L_1	L_2	TIE TOGETHER
LOW VOLTAGE	T_1 T_3	T_2 T_4	——————
HIGH VOLTAGE	T_1	T_4	T_2 T_3

FIGURE 21-9 Schematic diagram of a dual-voltage, repulsion-start, induction-run motor.

© Cengage Learning 2014

REPULSION-INDUCTION MOTOR

The operating characteristics of a repulsion-induction motor are similar to those of the repulsion-start, induction-run motor. However, the repulsion-induction motor has no centrifugal mechanism. It has the same type of armature and commutator as the repulsion motor, but it has a squirrel-cage winding beneath the slots of the armature.

Figure 21–10 shows a repulsion-induction motor armature with a squirrel-cage winding. One advantage of this type of motor is that it has no centrifugal device requiring maintenance. The repulsion-induction motor has a very good starting torque because it starts as a repulsion motor. At start-up, the repulsion winding predominates; however, as the motor speed increases, the squirrel-cage winding is used most. The transition from repulsion to induction operation is smooth because no switching device is used. In addition, the repulsion-induction motor has a fairly constant speed regulation from no load to full load because of the squirrel-cage

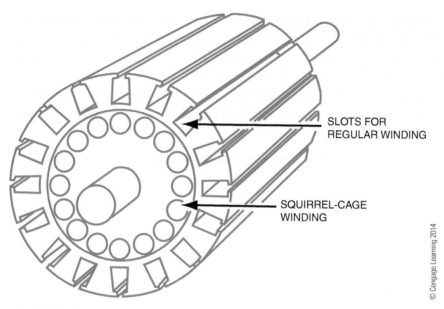

© Cengage Learning 2014

SLOTS FOR
REGULAR WINDING

SQUIRREL-CAGE
WINDING

FIGURE 21–10 An armature of a repulsion-induction motor.

winding. The torque-speed performance of a repulsion-induction motor is similar to that of a DC compound motor.

A repulsion-induction motor can be operated on either 115 volts or 230 volts. The stator winding has two sections connected in parallel for 115-volt operation and in series for 230-volt operation. The markings of the motor terminals and the connection arrangement of the leads are the same as in a repulsion-start, induction-run motor. The symbol in Figure 21–8 also represents a repulsion-induction motor (as well as a repulsion-start, induction-run motor and a repulsion motor).

NATIONAL ELECTRICAL CODE® REGULATIONS

National Electrical Code® requirements for the motor branch-circuit overcurrent protection, motor overcurrent protection, and wire sizes for motor circuits are provided in *Article 430* of the *Code.*

SUMMARY

AC series motors are conduction motors, just like series DC motors. The construction is slightly different because the magnetic field changes affect the inductance of the iron. The principle of operation is the same as that of the series DC motor. The armature keeps the same magnetic polarity of the rotor, reacting with the same magnetic field of the stator through the process of commutation.

Repulsion motors are available in three basic designs: (1) repulsion motors, (2) repulsion-start, induction-run motors, and (3) repulsion-induction motors. These motors are easy to recognize because they are AC induction motors but use a commutator and brushes. The important point to remember is that the motors have neutral positions of the brush mountings that yield no motor movement. These neutral positions are referred to as hard or soft neutral. The brushes are shifted off soft neutral to give the motor the desired direction of rotation.

ACHIEVEMENT REVIEW

A.   Completely answer the following questions.

1.   a. Describe the basic differences in construction between the concentrated-field and the distributed-field types of universal motors. b. Draw the schematic diagram for each type of motor.

a.   _____

_____

_____

_____

_____

_____

_____

b.

2. What is the function of the compensating winding in a two-field compensated universal motor? _____

_____

_____

_____

3. Describe three methods of controlling the speed of universal motors._____

_____

_____

_____

_____

_____

_____

4. Why does a universal motor spark excessively at the commutator if its direction of rotation is reversed? _____

_____

_____

_____

_____

5. A DC series motor operates unsatisfactorily on AC. What are the primary reasons for this fact? _____

_____

_____

_____

6. What is a repulsion motor, and how is rotation produced?_____

_____

_____

_____

_____

_____

_____

7. Name one application of a repulsion motor._____

_____

_____

8. Describe the operation of a repulsion-start, induction-run motor. _____

_____

_____

_____

9. Explain the difference between the brush-lifting type of repulsion-start, induction-run
   motor and the brush-riding type of repulsion-start, induction-run motor.

_____

_____

_____

_____

_____

10. A 2 hp, 230-volt, 12-ampere, single-phase repulsion-start, induction-run motor is con-
    nected directly across the rated line voltage.
    a. Determine the overcurrent protection for the branch circuit feeding this motor.

_____

_____

_____

_____

_____

   b. Determine the running overcurrent protection for this motor.

   _____

   _____

   _____

12. Describe the construction of a repulsion-induction motor.

   _____

   _____

   _____

   _____

   _____

13. What is one advantage to the use of the repulsion-induction motor as compared to the repulsion-start, induction-run motor? _____

   _____

   _____

14. Explain how the direction of rotation is changed for any one of the three types of single-phase repulsion motors covered in this unit. _____

   _____

   _____

   _____

   _____

B. Select the correct answer for each of the following statements, and place the corresponding letter in the space provided.

15. The operation of a DC shunt motor from an AC source is impractical because _____
   a. too much torque is developed at start-up.
   b. the starting current is too high.
   c. the shunt field inductance is too high.
   d. the shunt field inductance is too low.

16. A series DC motor fails to operate satisfactorily on AC due to _____
    a. eddy currents and high field voltage drop.
    b. excessive heat and low field voltage drop.
    c. low reactance of the armature and field.
    d. high armature reluctance and low field reactance.

17. The frames of universal motors are made of _____
    a. rolled steel.
    b. cast iron.
    c. aluminum.
    d. all of the above

18. A compensating winding _____
    a. increases the reactance in the armature on AC.
    b. reduces the reactance in the armature on AC.
    c. reduces the reactance in the armature on DC.
    d. increases the reactance in the armature on DC.

19. After changing the direction of rotation of a universal motor, the_____
    a. brushes must be rotated for sparkless commutation.
    b. field connections must be shifted.
    c. field reactance must be decreased.
    d. field reactance must be increased.

20. Insert the correct word or phrase to complete each of the following statements.
    a. A repulsion-induction motor has a good _____ and a fairly good

       _____.

    b. A repulsion motor has a high starting torque and its speed rapidly decreases with

       _____.

    c. The centrifugal short-circuiting device on a repulsion-start, induction-run motor
       operates at approximately _____ of the rated speed.
    d. Both the repulsion-start, induction-run motor and the repulsion-induction motor
       operate as _____ after
       they have accelerated to rated speed.

UNIT 22

ENERGY-EFFICIENT MOTORS

OBJECTIVES

After studying this unit, the student should be able to

- determine the operating characteristics of an energy-efficient motor.

- select the proper efficiency points for motors operating at reduced frequencies.

- determine when to use inverter duty motors.

- select integrated motor and controller applications.

Standard motor efficiency is calculated by the simple formula of watts output of the motor divided by watts input. Watts output is calculated by determining the mechanical horsepower output and multiplying by 746 watts. Each output horsepower of mechanical energy is equal to 746 watts of electrical energy. The input wattage is measured by an electrical wattmeter, either single phase or three phase, or it is calculated by making current and voltage measurements and power factor measurements (see Figure 22–1).

The following formulas provide the input watts to an AC motor.

For single-phase motors, use the formula:

Watts input = line volts × line current × % power factor

For three-phase motors, use the formula:

Watts input = line volts × line current × 1.73 × % power factor

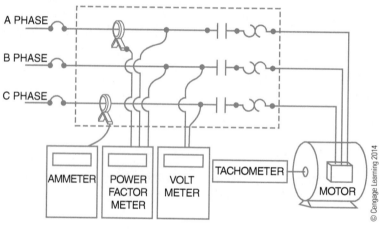

FIGURE 22–1 Power measurement for input to motor.

Calculating the output power is sometimes difficult. Horsepower is a function of the motor torque, the speed of the shaft, and the amount of time the motor operates. This means that the motor can move a weight (measured in pounds) a distance (measured in feet) over time (measured in minutes). Originally, a single horsepower was calculated as the amount of effort required to move 33,000 pounds 1 ft in 1 minute. In other words 1 hp = 33,000 ft-lbs/min. To measure the output, the force on the shaft (ft-lbs) and the RPM (how far a weight at some distance from the center of the shaft would spin in one minute) must be measured. If the motor can be connected to a test device, called a *dynamometer*, the output horsepower can be directly measured. If the motor can be connected to a device called a *prony brake*, the torque can be calculated. The torque in foot-pounds is then used in the horsepower formula: hp = 2 × π × torque × RPM / 33,000. If the motor is connected to a driven load, horsepower measurement is not an easy task. Output power is often estimated using various techniques.

Load estimation can help determine how much actual work the motor is doing, measured in horsepower. This is compared to the input watts to determine how efficient the motor is at converting electrical energy to mechanical energy. One method is to use power draw comparisons. If the nameplate efficiency is used as a guide to full-load efficiency, then a partial load can be determined by using the following formula:

$$\text{output at partial load} = \text{measured input watts/full-load input watts}$$

Full-load input watts are calculated by knowing the rated horsepower multiplied by 746, divided by the rated efficiency. This assumes that a partial load operates at nearly the same efficiency as a full load. This is not always the case, and other methods must be used to approximate the actual load on the motor.

Line current measurement can give another approximation of motor performance. The line current draw of the motor is roughly proportional to the actual load until the motor is under 50% loaded; then, the ratio is no longer true. For example, if the actual motor current is 80% of the full-load value, the motor is producing roughly 80% of the rated horsepower output.

If you can measure the operating speed of a motor under load with a tachometer, as in Figure 22–2, then the slip method can be used. Use the RPM slip divided by the result of synchronous RPM minus full-load RPM. Use this answer to multiply the nameplate hp to find the running hp.

FIGURE 22–2 Tachometer.

© Cengage Learning 2014

● *Example:* Measured RPM is 1770 at partial load. Full-load nameplate RPM is 1740, and the nameplate horsepower is 10 hp. Partial-load slip is 30 RPM (1800 – 1770).

$$\frac{30}{(1800 - 1740)} = 0.5$$

The motor is operating at 0.5 × 10 hp, or 5 hp output.

MOTOR LOSSES

Motor losses within the motor affect motor efficiency. In the conversion of energy from electrical to mechanical, some losses are to be expected. Motor losses are grouped into several categories.

Core losses in the motor are made up of losses in the iron core. These losses are due to eddy currents in the motor iron, and hysteresis losses in the magnetic circuit of the motor iron. To reduce hysteresis losses, high-grade steel with a higher silicon content is used to reduce the friction of the magnetic materials and therefore reduce the heat produced in changing the magnetic fields of the motor iron. To reduce the eddy currents, the laminations are sliced thinner and electrically insulated from one another. In addition, by lengthening the core, there is less flux density in the core material and reduced losses due to magnetic effects.

Friction and windage losses are the results of the motor bearings and the friction encountered by the spinning rotor. If the friction of the bearings can be reduced, the motor will have less opposition to the turning shaft and reduced losses within the motor. Windage is the result of the needed cooling of the motor windings. If other losses, including core and copper losses, can be reduced, there is less need for cooling and the fan size and air movement can be reduced, which again reduce losses within the motor.

Copper losses include stator losses that occur in the copper winding of the stator. They are referred to as I^2R losses. If the wire size in the stator winding can be increased, then the R (resistance) is reduced and the losses are reduced. The wire size is increased by larger slots in the stator and reduced thickness of the wire insulation. Copper losses occur in the rotor winding as well. By reducing the resistance of the rotor, the same benefits of reducing the wattage losses in the stator are realized, and motor efficiency is increased.

Stray losses make up the remaining motor losses. Stray losses occur because the magnetic flux created in the stator is lost and never reacts with the rotor conductors to produce torque. By lengthening the core, stray losses are reduced. To accomplish all these reductions in losses, redesigns were needed. In most cases, more expensive materials and processes are used in high-efficiency motors. The original purchase price of a high-efficiency motor is higher than the price of a comparable horsepower, standard-efficiency motor. The operating cost for electricity for a high-efficiency or a premium-efficiency motor can be less than that for a standard motor if the motors are properly selected to take advantage of the actual load conditions of the motor.

MOTOR EFFICIENCY DESIGNATIONS

Premium efficiency motors are designed to run at 94.1% efficiency at full load. A high-efficiency motor is designed to run at 91.0% efficiency. If a motor is more efficient, it means that a higher percentage of the input power measured in electrical watts is converted to output mechanical horsepower, compared to a lower efficiency motor. All electric motors have losses as described previously. The motor manufacturers have worked to reduce these internal losses even more producing a "premium efficiency" motor. If you were to run a "premium efficiency" 25 HP motor at full load for 24 hours per day and 7 days per week (8736 hours) at a total cost of $0.10 per kilowatt hour, you would pay:

- "Premium efficiency 94.1% eff" motor cost to operate: $(\frac{25 \text{ hp} \times 746 \text{ W}}{.941 \times 1000}) \times 8736$ hours $\times$ $0.10 per kilowatt hour $= (19.81 \text{ kW}) \times 8736 \times 0.10 = \$17,314$

- A "high-efficiency 91% eff" motor would cost : $(\frac{25 \text{ hp} \times 746}{.910 \times 1000}) \times 8736 \times .10 =$ $(20.5 \text{ kW}) \times 8736 \times 0.10 = (20.5 \text{ kW}) \times 8736 \times 0.10 = \$ 17,904.$

This represents a savings of approximately $590.00 per year.

A comparison formula:

Savings in dollars $= (0.746 \text{ kW per hp}) \times (\text{hp}) \times (\frac{100}{lower \text{ } eff \%} - \frac{100}{high \text{ } eff \%}) \times (RT)$ hours of run time per year $\times$ (E) energy cost in dollars per kWh, or approximately $586.00

Typically a premium efficiency motor runs at a cooler temperature, which extends motor life. The insulation temperature is a contributing factor to motor life. It is estimated that the insulation life doubles for every 10°C (50°F) reduction in operating temperature. A cooler temperature also aids in bearing life. The cooler the motor runs usually means the longer the expected life.

Electric motors consume nearly 60% of the electric energy in the United States; therefore, the Energy Independence and Security Act (EISA) was created in 2007. The actual improvements in energy efficiencies were actually realized in 2010 by the motor manufacturers.

Motor manufacturers list their premium efficiency motors under different names. Typically, the larger the motor hp, the higher the efficiency ratings. The highest efficiency ratings appear between 75% and 100% load. Power factor also typically improves with the hp rating of the motor and again tends to peak about 75% to 100% of full load.

Stator losses traditionally made up about 66% of the power losses in the I^2R losses of the stator windings. To increase efficiency, more copper was added to the stator windings to reduce the R and therefore reduce the I^2R losses. To reduce rotor losses, the slip was decreased to change the way the stator flux reacted with the rotor. This required increased conduction of the rotor bars. The following chart compares a standard efficiency motor to a premium efficiency motor and estimates the payback period for the higher priced premium motor. The comparisons are made at 75% of full load and use 7.5 cents per kWH for comparison purposes. The first chart compares replacing a functioning motor with a higher efficiency motor just to gain operating energy savings.

Energy and Cost Savings Available When Replacing Serviceable Standard Efficiency Motor with an EPAct-level or NEMA Premium Motor							
	Std Efficiency Motors, Average Efficiency		Replace with EPAct Motors				
HP	%Eff. at 75% load	Annual Energy Use (kWh), cost	Purchase Price (35% disc)	% Eff. at 75% load	Annual Energy Use (kWh), cost	Annual Saving kWh, $	Payback Period
5	84.0	26,644	$233	88.2	25,374	1,270	2.44
		$1,988			$1,903	$95	
10	86.75	51,653	$375	90.0	49,773	1,919	2.60
		$3,874			$3,730	$144	
15	87.55	76,771	$562	91.0	73,780	2,991	2.50
		$5,758			$5,534	$224	
20	89.3	100,206	$666	92.6	96,626	3,579	2.48
		$7,515			$7,247	$268	
25	89.9	124,457	$800	93.1	119,952	4,505	2.36
		$9,334			$8,996	$338	
50	91.6	244,211	$1617	93.9	238,027	6,185	3.48
		$18,316			$17,852	$464	

The second chart compares replacing a failed motor with a premium efficiency motor and compares the payback time for the difference in motor cost and operating expenses.

	EPAct Motors		Motor	NEMA Premium Motors		
HP	% Eff. At 75% load	Annual Energy Use (kWh), cost	Purchase Cost Premium	Annual Energy Ese (kWh), cost	Annual Savings, kWh, $	Payback Period
5	88.2	25,374	$70	24,729	1,855	1.43
		$1,903		$1,855	$48	
10	90.0	49,773	$143	48,547	1,187	1.60
		$3,730		$3,641	$89	
15	91.0	73,780	$115	72,505	1,275	1.20
		$5,534		$5,438	$96	
20	92.6	97,030	$158	95,846	1,185	1.77
		$7,277		$7,188	$89	
25	93.1	120,248	$265	119,043	1,205	2.93
		$9,019		$8,928	$90	
50	93.9	238,316	$177	235,331	2,985	2.64
		$17,874		$17,650	$224	

Comparison of Annual Savings and Simple Payback When Comparing Replacement of a Failed Motor with an Epact-level or NEMA Premium Motor

Note: MototMaster may produce apparent small mathematical errors due to rounding

© Cengage Learning 2014

Several other examples, worked out using MotorMaster+ software, can be found elsewhere on this Web site.

MOTOR ENERGY REQUIREMENTS

When selecting motors to drive loads, an awareness of the loading effects on the energy drawn by the motor is important. The effects on the motor input energy can be estimated by how the motor is used and at what speed it is driven. The physical laws that determine the operating characteristics and application are known as the *affinity laws* or the *cube load laws*. In short, the laws of energy consumption state that the energy required to drive fans or pumps (motors used for moving volumes of air or water) is proportional to the cube of the speed. For example, if the speed of a fan is reduced to 50% of the rated full-load speed, the fan slows to 50% of the RPM. This in turn reduces the air volume delivered to 50% of the full volume. The change is a linear proportion. The air pressure also changes as the square of the speed changes. For 50% of normal speed, the pressure drops to 0.5^2, or 25% of normal full-speed pressure. However, the energy consumption changes to 50% cubed, or 0.5^3, yielding only 12.5 % of the energy used.

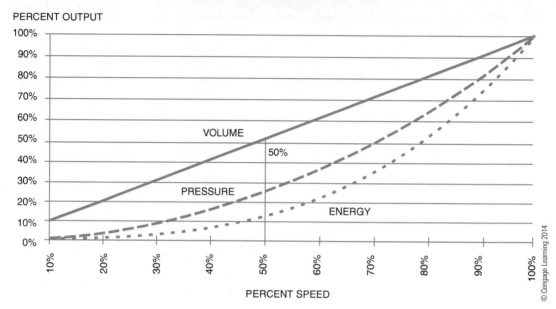

FIGURE 22-3 Affinity laws, also known as cube load laws.

Another example is that a reduction to 80% of the full-speed condition reduces the air volume to 80% and the pressure to 64% of the full pressure, but the energy used is reduced to 0.80 cubed, or 51.2% of the input electrical energy (see Figure 22-3).

If the fan volume and pressure are sufficient to supply the needs of the facility, then operating below normal speed results in large savings for large fans and pumps. Variable frequency drives are ideal for tailoring the actual operating speed to facility needs. A typical application is an office that needs less volume and pressure for environmental air during off hours. A simple program to reduce speed may yield large energy and dollar savings.

Pumps work on the same physical laws when the speed of the pump, but not the speed of the impeller, is being changed. If the pressure and volume of a reduced-speed pump meets the demand of the mechanical system, then reduced speeds may be an effective energy reduction technique.

Many premium efficiency motors are also designed to be inverter duty rated. These motors are designed to run with variable frequency drives without affecting the life of the motor.

When using energy reduction techniques, such as variable frequency drives, it may be wise to use motors designed for use with inverter drives. Inverter-duty and vector-drive motors are designed for this purpose. Because PWM and vector drives are switching the power at high speed, there is a tendency for a standard motor's magnet wire (motor winding) insulation to break down

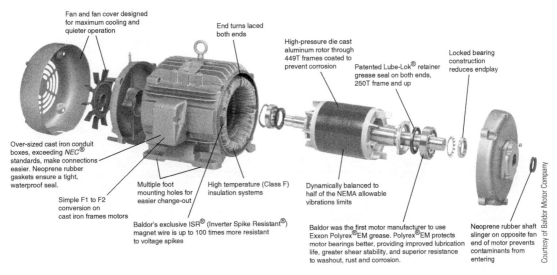

Fan and fan cover designed
for maximum cooling and
quieter operation

End turns laced
both ends

High-pressure die cast
aluminum rotor through
449T frames coated to
prevent corrosion

Patented Lube-Lok® retainer
grease seal on both ends,
250T frame and up

Locked bearing
construction
reduces endplay

Over-sized cast iron conduit
boxes, exceeding NEC®
standards, make connections
easier. Neoprene rubber
gaskets ensure a tight,
waterproof seal.

Multiple foot
mounting holes for
easier change-out

High temperature (Class F)
insulation systems

Dynamically balanced to
half of the NEMA allowable
vibrations limits

Simple F1 to F2
conversion on
cast iron frames motors

Baldor's exclusive ISR® (Inverter Spike Resistant®)
magnet wire is up to 100 times more resistant
to voltage spikes

Baldor was the first motor manufacturer to use
Exxon Polyrex®EM grease. Polyrex®EM protects
motor bearings better, providing improved lubrication
life, greater shear stability, and superior resistance
to washout, rust and corrosion.

Neoprene rubber shaft
slinger on opposite fan
end of motor prevents
contaminants from
entering

Courtesy of Baldor Motor Company

FIGURE 22-4 Totally enclosed, blower-cooled (TEBC) inverter motor.

because of spikes that occur and punch through the standard wire insulation. Some motor manufacturers produce a line of motors specifically designed with magnet wire resistant to transient spikes (see Figure 22-4). Often these motors need different cooling techniques that are standard for totally enclosed, fan-cooled (TEFC) motors. TEFC motors have external fins on the motor that radiate the heat to the motor surface. An external fan then blows ambient air over the fins to cool the motor. The fan is attached to the motor shaft and runs at the same speed as the motor shaft. If the motor uses a higher temperature–rated insulation on the windings, such as class H insulation, the motor can be totally enclosed, nonventilated (TENV) or totally enclosed, blower cooled (TEBC).

Along with changes in motor design to accommodate different or lower than normal speeds, motor wiring also deserves attention. The wires that carry the power from the PWM drive to the motor must be able to handle the normal power levels of the PWM pulses, as well as the standing waves that produce high voltages on the wire and the insulation. It has been determined that there may be corona breakdown of the insulation caused by these high voltages. Electrical noise has also been an observed problem for motors driven by variable frequency drives (VFDs). Cable manufacturers have developed products to counteract the effects of the problems associated with long lengths of leads from the VFD to the motor. Figure 22-5 shows examples of special inverter-driven motor leads.

Another change in the motor and drive industry is the integration of the motor and the drive in a single unit (see Figure 22-6). The drive has been specifically matched to the specific inverter-duty motor, which is a TEBC design. The internal leads from the drive to motor are designed for long life because of the reduction of corona and the reduction of reflected voltage wave. This integrated solution provides for perfectly matched components and easy, one-step, no-engineering installation.

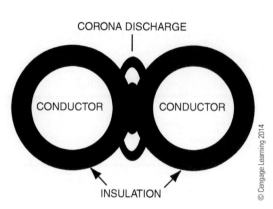

FIGURE 22-5 Example of new cable design to help eliminate problems with wiring supplying VFDs.

FIGURE 22-6 Integrated electronic drive and motor combination.

Summary

Energy-efficient motors are designed to operate with less slip than conventional motors and to better use the input wattage to create output wattage or horsepower. In design and manufacturing, these motors are made with higher grade silicon steel and thinner laminations of the core steel to reduce core losses. The magnet wire used in the windings is larger gauge to reduce copper losses. If less heat is produced (lost) within the motor, the fan can be smaller, and less windage loss is incurred. If the bearings are of better quality than standard duty, then less friction is developed, and reduced losses occur there. In all, every effort is made to reduce the wattage lost inside the motor in the conversion of electrical to mechanical energy. To estimate the losses and efficiency, the output load must be determined. Several methods are used to estimate the load, and a few methods used to actually measure the output load.

When reducing the energy requirements of a fluid drive system—air or liquid—the results of the cube load energy law are effective. This law relates the effect on the fluid system to the amount of input energy needed to drive the load. To reduce the speed of a drive system to conserve energy, a VFD is used. Other factors, such as adequate cooling of the motor at lower speed, must be considered. Standard motors and supply leads used with PWM drives often fail due to voltage reactions. Motors are available with inverter-duty windings that are used to counteract the effects of the electronic drives. Leads are available to counteract the effects of the electronic drives on the motor supply leads. All these solutions can be purchased as a single matched unit in an integrated drive and motor set.

ACHIEVEMENT REVIEW

1. How many watts are equivalent to 10 hp?_____

2. Show the formula for determining the efficiency of a single-phase AC motor.

    _____

3. Calculate the horsepower output of a motor operating at 1750 RPM at 60 Hz. The full-load nameplate RPM is 1725, and the rated horsepower is 7.5 hp.

    _____

    _____

    _____

    _____

4. Explain how the eddy current losses in a motor are reduced.

    _____

    _____

5. Name the two parts of the motor that create copper losses.

    _____

    _____

6. If the motor speed for a fan drive is reduced to 60% of the design speed, the motor power will be reduced to approximately _____ of the rated power input.
    a. 21.5%
    b. 36%
    c. 60%
    d. 64%

7. What does TEBC stand for on a motor nameplate?

    _____

    _____

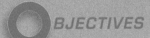

OBJECTIVES

After studying this unit, the student should be able to

- determine, for several types of three-phase AC induction motors, the
  - size of the conductors required for three-phase, three-wire branch circuits.
  - sizes of fuses used to provide starting protection.
  - disconnecting means required for the motor type.
  - size of the thermal overload units required for running overcurrent protection.
  - size of the main feeder to a motor installation.
  - overcurrent protection required for the main feeder.
  - main disconnecting means for the motor installation.
- use the *National Electrical Code® (NEC®)*.

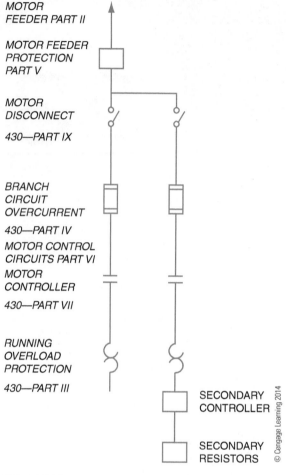

MOTOR
FEEDER PART II

MOTOR FEEDER
PROTECTION
PART V

MOTOR
DISCONNECT

430—PART IX

BRANCH
CIRCUIT
OVERCURRENT

430—PART IV
MOTOR CONTROL
CIRCUITS PART VI
MOTOR
CONTROLLER

430—PART VII

RUNNING
OVERLOAD
PROTECTION

430—PART III

SECONDARY
CONTROLLER

SECONDARY
RESISTORS

© Cengage Learning 2014

FIGURE 23–1 Line diagram of motor control system.

The work of the modern electrician requires a knowledge of *National Electrical Code®* requirements that govern three-phase motor installations and the ability to apply these requirements to installations. The elements of a motor circuit are shown in Figure 23–1.

This unit outlines the procedure for determining the wire size and the proper overload and starting protection for a typical three-phase motor installation. The motor installation example consists of a feeder circuit feeding three branch circuits. Each of the three branch circuits is connected to a three-phase motor of a specified horsepower rating. The feeder circuit and the branch circuits have the necessary overcurrent protection required by the *NEC®*.

THREE-PHASE MOTOR LOAD

The industrial motor installation described in this example is connected to a 230-volt, three-phase, three-wire service (Figure 23–2). The load of this system consists of the following branch circuits.

- One branch circuit that feeds a three-phase AC induction motor rated at 230 volts, 28 amperes, 10 hp, with a code letter F marking.

- One branch circuit that feeds a three-phase AC induction motor rated at 230 volts, 64 amperes, 25 hp, with a code letter B marking.

- One branch circuit that feeds a wound-rotor induction motor rated at 230 volts, 54 amperes, and 20 hp. The full-load rotor current is 60 amperes.

BRANCH CIRCUIT FOR EACH MOTOR

The values given in *NEC® Table 310.15(B)(16)* shall be used with the *Code* book current for motors in determining ampere capacity (ampacity) of the conductor according to *Article 430.6.*

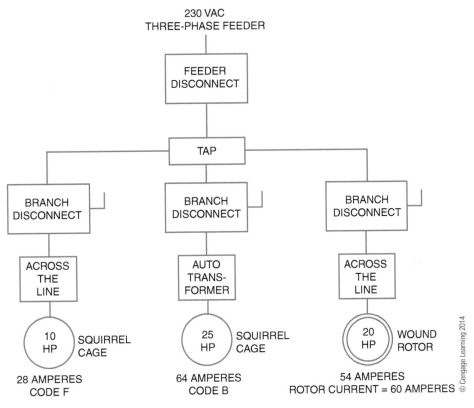

FIGURE 23-2 Branch circuit for each motor.

Three specific facts must be determined for each of the three branch circuits constituting the load of the installation:

- The size of the conductors for each three-phase, three-wire branch circuit.
- The fuse size to be used for short-circuit protection. The fuses protect the wiring and the motor from any faults or short circuits in the wiring or motor windings.
- The size of the thermal overload units to be used for running protection. The overload units protect the motor from potential damage due to a continued overload on the motor.

Note: The full-load amperes shall be taken from the motor's nameplate only for calculating thermal overload units. [See NEC® Article 430.6(A)(2).] Other calculations are based on *Code-rated values from Articles 430.248, 430.249,* and *430.250.* Where motor system disconnecting means and controllers are determined from horsepower, voltage rating, and design letter, then *Table 430.251(B)* is used.

BRANCH CIRCUIT 1

The first branch circuit feeds a three-phase AC induction motor. The nameplate data of this motor follow:

Squirrel-Cage Induction Motor	
Volts 230	Amperes 28
3 Phase	Speed 1735 RPM
Code Letter F	Frequency 60 Hertz
10 Horsepower	Temperature Rating 40° (104°) Celsius

© Cengage Learning 2014

Conductor Size

NEC® 430.22 states that branch-circuit conductors supplying a single motor shall have a carrying capacity equal to not less than 125% of the full-load current rating of a motor. This general rule may be modified according to *430.22(A–G)*.

The following procedure is used to determine the size of the conductors of the branch circuit feeding the 10 hp motor:

1. The 10 hp motor has a full-load current rating of 28 amperes. According to *Table 430.250:*

$$28 \times 125\% = 35 \text{ amperes}$$

2. Using 35 amperes and referring to *Table 310.15(B)(16)*, a proper size of conductor is selected. This process requires the electrician to determine the temperature ratings of each termination used and the ampere rating of the equipment circuit. According to *NEC® Article 110.14(C)*, the temperature rating of the conductor used to determine the ampacity must not exceed the temperature rating of any of the connections. Unless all the terminations are marked for a higher temperature, the column in *310.15(B)(16)* marked 60°C is selected to determine the conductor ampacity. Even if using a standard building wire type THHN, the conductor size is 8 AWG in the 60°C column.

   If all the terminations in the branch circuit are rated for 75°C, then the second column in *310.15(B)(16)* can be used for all wire ampacities. *Article 110.14(C)(1)(a)(4)* states that motors with design letters B, C, or D may use the 75°C rating for terminations and wire ampacity. If all the other terminations in that circuit are rated at 75°C, then 10 AWG—75°C or 10 AWG—90°C wire may be used.

3. *Table C1 in NEC Annex C* indicates that three 8 AWG THHN conductors will fit in a trade size 1/2 EMT conduit (metric designator 16).

The squirrel-cage induction motor is to be connected directly across the rated line voltage through an ATL motor starter. For this example, the branch-circuit, short-circuit, and ground-fault protection for this motor consists of three standard nontime-delay fuses enclosed in a safety switch located on the line side of the magnetic starter. According to *430.109(A)* or as specified in *430.109(B–G)* of the *Code*, this switch shall be a motor-circuit switch with a horse-power rating, a circuit breaker, or a molded case switch, or shall be a listed device.

Motor Branch-Circuit, Short-Circuit, and Ground-Fault Protection

The branch-circuit, short-circuit, and ground-fault protection for a three-phase AC induction motor is provided in *Table 430.52*. For the branch circuit 1 motor being considered, the motor circuit overcurrent device shall not exceed 300% of the full-load current of the motor (nontime-delay fuses). *Article 430.52* with exceptions applies to *Table 430.52*.

The branch-circuit fuse protection for the branch circuit feeding the squirrel-cage motor is a *squirrel cage*—other than Design B energy efficient.

Because the 10 hp motor has a full-load current rating of 28 amperes, and given the appropriate value from *Table 430.52* for a nontime-delay fuse, then

$$28 \times 300\% = 84 \text{ amperes} / 90 \text{ amperes}$$

NEC® Article 430.52 Exception 1 states that if the values for branch-circuit protective devices as determined using the percentages in *Table 430.52* do not correspond to the standard device sizes or ratings, then the next larger size rating or setting should be used.

NEC® 240.6 indicates that the next larger, standard-size fuse above 84 amperes is 90 amperes. Standard nontime-delay cartridge fuses rated at 90 amperes may be used as the branch-circuit protection for this motor circuit.

The branch-circuit, short-circuit, and ground-fault protection may also be calculated using a time-delay fuse. From *Table 430.52*, the second column is selected and 175% of 28 amperes is calculated ($1.75 \times 28 = 49$ amps). The next larger size is used. In this example, 50-ampere fuses would be the choice. The *Code* allows the electrician to increase the size of the fuse according to the exceptions in *430.52(C)(1)*.

Disconnecting Means

According to the table for safety switches (Figure 23–3), the disconnecting means for this 10 hp motor is a 15 hp, 100-ampere–rated safety switch in which the 90-ampere fuses are installed.

Because these safety switches are dual rated, it is permissible to install a 60-ampere safety switch with a maximum rating of 15 hp if the time-delay fuses are appropriate for the starting characteristics of the motor. The size of the time-delay fuses installed in the safety switch depends on the degree of protection desired and the type of service required of the motor.

Three-Pole, Three-Fuse, 230-Volt AC Safety Switches		
	Approximate Manufacturer Horsepower Ratings	
Amperes	Standard	Maximum
30	3	7 1/2 *
60	7 1/2	15 *
100	15	30 *
200	25	60 *
400	50	100 *

* The *Electrical Construction Materials List* by Underwriters Laboratory, Inc. states that "some enclosed switches have dual horsepower ratings, the larger of which is based on the use of fuses with time delay appropriate for the starting characteristics of the motor. Switches with such horsepower ratings are marked to indicate this limitation and are tested at the larger of the two ratings."

© Cengage Learning 2014

FIGURE 23-3 Table for safety switches.

Time-delay fuses ranging in size from 35 amperes to 60 amperes may be installed in the safety switch. Refer to *Article 430, Part IX,* for requirements of motor and controller disconnecting means.

Running Overload Protection

Running overload protection consists of three current monitors, housed in the ATL motor starter. (See the *Exception* following *NEC® Table 430.37* for an exception to this statement.)

 NEC® 430.32(A)(1) states that the running overload protection (motor and branch-circuit overload protection) for a motor shall trip at not more than 125% of the full-load current (as shown on the nameplate) for motors with a marked temperature rise 40°C (104°) or less. In this example, the nameplate is the same as the code value, which is uncommon.

 The trip current of the thermal units used as running overcurrent protection is

$$28 \times 125\% = 35 \text{ amperes}$$

 When the selected overload setting of the relay is not sufficient to start the motor or to carry the load, *430.32(C)* permits the use of the next higher size or rating, but must trip at no more than 140% of the full-load motor current.

BRANCH CIRCUIT 2

A second branch circuit feeds a three-phase AC induction motor. The nameplate data for this motor follow.

Squirrel-Cage Induction Motor	
Volts 230	Amperes 64
3 Phase	Speed 1740 RPM
Code Letter B	Frequency 60 Hertz
25 Horsepower	Temperature Rating 40° Celsius

© Cengage Learning 2014

Conductor Size

The following procedure is used to determine the size of the conductors of the branch circuit feeding the 25 hp motor:

1. The 25 hp motor has a full-load current rating of 68 amperes (see *NEC®, Table 430.250*). (According to the *Code, 430.22*, 125% is needed for ampacity.)

$$68 \times 125\% = 85 \text{ amperes}$$

2. *Table 310.15(B)(16)* indicates a 3 AWG Type TW or THHN copper conductor or a 3 Type THW conductor. (Assume 60°C terminations.)

3. *Table C1* of *Annex C* shows that three 3 TW, THW, or THHN or THW conductors may be installed in a trade size 1-in. conduit.

Note: *NEC®* 300.4G requires that where conductors of 4 size or larger enter an enclosure, an insulating bushing or equivalent must be installed on the conduit.

Motor Branch-Circuit Protection

The 25 hp, squirrel-cage induction motor is to be started using an autotransformer. The branch-circuit overcurrent protection for this motor circuit consists of three nontime delay fuses located in a safety switch mounted on the line side of the starting compensator (autotransformer).

For a squirrel-cage induction motor that is not Design B, *NEC® Table 430.52* requires that the branch-circuit overcurrent protection not exceed 300% of the full-load current of the motor. Because the 25 hp motor has a full-load current rating of 68 amperes (*NEC Table 430.250*), the branch-circuit overcurrent protection for the branch circuit feeding this motor is

$$68 \times 300\% = 204 \text{ amperes}$$

NEC® 240.6 does not show 204 amperes as a standard size for a fuse. However, 430.52 permits the use of a fuse of the next higher size if the calculated size is not a standard size. In this case, 200 amperes should be attempted. Therefore, three 200-ampere, nontime-delay fuses can be used as the branch-circuit protection for this motor. *Article 430.52, Exception 2,* allows the fuse to be increased, but not over 400%.

Disconnecting Means

According to the table for safety switches in Figure 23–3, the disconnecting means for the 25 hp motor is a 25 hp, 200-ampere safety switch in which the 200-ampere fuses are installed.

Time-delay fuses may be installed in safety switches. In this example, 175% × 68 amperes = 119 amperes. The next largest size is 125 fuses, and these may be used according to exceptions in 430.52. The safety switch would be the same size.

Running Overcurrent Protection (Motor Overload Protection)

The running overcurrent protection consists of three magnetic overloads located in the starting compensator. According to the nameplate, the motor has a full-load current rating of 64 amperes. The current setting of the magnetic overload units is set to trip at

$$64 \times 125\% = 80 \text{ amperes (trip current)}$$

BRANCH CIRCUIT 3

A third branch circuit feeds a wound-rotor induction motor. The nameplate data for this motor follow.

Wound-Rotor Induction Motor	
Volts 230	Stator Amperes 54
3 Phase	Rotor Amperes 60
20 Horsepower	Frequency 60 Hertz
Temperature Rating 40° Celsius	

© Cengage Learning 2014

Conductor Size (Stator)

The following procedure is used to determine the size of the conductors of the branch circuit feeding the 20 hp motor.

1. The 20 hp motor has a full-load current rating of 54 amperes. According to the NEC® 430.22 and *Table 430.250.*

$$54 \times 125\% = 67.5 \text{ amperes}$$

2. *Table 310.15(B)(16)* indicates an AWG 4 Type TW, THW, THHN conductor (70 amperes).

3. *Table C1* of *Annex C* shows three AWG 4 TW, THW, or THHN conductors may be installed in a trade size 1 in. conduit (metric designator 27).

Note: *Article 300.4(G)* requires that where conductors of 4 size or larger enter an enclosure, an insulating bushing or equivalent must be installed on the conduit.

Motor Branch-Circuit Protection

The 20 hp, wound-rotor induction motor is to be started by means of an ATL magnetic motor switch. This motor starter applies the rated three-phase voltage to the stator winding. Speed control is provided by a controller used in the rotor or secondary circuit. All resistance of the controller is inserted in the rotor circuit when the motor is started. As a result, the inrush starting current to the motor is less than if the motor were started at full voltage.

The branch-circuit, short-circuit protection of a wound-rotor induction motor is required by *Table 430.52* of the *Code* not to exceed 150% of the full-load running current of the motor. This is for nontime-delay and time-delay fuses.

The full-load current equals 54 amperes for a 20 hp wound-rotor motor. The branch-circuit overcurrent protection for the branch circuit feeding this motor is

$$54 \times 150\% = 81 \text{ amperes}$$

NEC® 240.6 does not show 81 amperes as a standard fuse size. *Article 430.52* allows the next larger size. A 90 ampere fuse should be chosen.

Disconnecting Means

According to the table for safety switches in Figure 23–3, the disconnecting means for the 20 hp motor is a standard 25 hp, 200-ampere safety switch. Reducers must be installed in this switch to accommodate the 90-ampere fuses required for the motor branch-circuit protection. Because of the dual rating of these safety switches, it is permissible to use a 100-ampere switch with a maximum rating of 30 hp. In this case, standard 90-ampere, nontime-delay fuses or 90-ampere, time-delay fuses may be installed.

Running Overcurrent Protection (Motor Overload Protection)

The running overcurrent protection consists of three thermal overload units located in the ATL magnetic motor starter (except as indicated in the note following *Table 430.37*). According to the nameplate, the motor has a full-load current rating of 54 amperes. The rated trip current of each thermal unit is

$$54 \times 125\% = 67.5 \text{ amperes}$$

Conductor Size (Rotor)

The rotor winding of the 20 hp, wound-rotor induction motor is rated at 60 amperes. The following procedure is used to determine the size of the conductors for the secondary circuit from the rotor slip rings to the drum controller:

1. *NEC*® *430.23(A)* requires that the conductors connecting the secondary of a wound-rotor induction motor to its controller have a current-carrying capacity not less than 125% of the full-load secondary current of the motor for continuous duty.

$$60 \times 125\% = 75 \text{ amperes}$$

2. *Table 310.15(B)(16)* indicates that several types of copper conductors can be used: AWG 3 Type TW, Type THW, or Type THHN, assuming 60° terminations.

3. *Table C1* of *Annex C* shows that A trade size 1 in. conduit is required for three 3 THHN, TW or THW wires.

Note: *Article 300.4(G)* requires the use of insulating bushings or the equivalent on all conduits containing conductors of 4 size or larger entering enclosures.

If the resistors are mounted outside the speed controller, the current capacity of the conductors between the controller and the resistors shall be not less than the values given in *Table 430.23(C)*.

For example, the manual speed controller used with the 20 hp, wound-rotor induction motor is to be used for heavy intermittent duty. *NEC*® *430.23(C)* requires that the conductors connecting the resistors to the speed controller have an ampacity not less than indicated in *Table 430.23(C)*. This is between 35% and 110%, based on the duty of the resistors. For example, if we use heavy intermittent duty, then 85% is used.

$$60 \times 85\% = 51 \text{ amperes}$$

Table 310.15(B)(16) indicates that 51 amperes can be carried safely by AWG 6 wire. As a result, the temperatures generated at the resistor location are an important consideration.

NEC® *430.32(E)* states that the secondary circuits of wound-rotor induction motors, including the conductors, controllers, and resistors, shall be considered as protected against overload by the motor running overcurrent protection in the primary or stator circuits, Therefore, no further overcurrent protection is necessary in the secondary rotor circuit.

MAIN FEEDER

When the conductors of a feeder supply two or more motors, the required wire size is determined using *Code* rules. *Article 430, Part II, 430.24* states that the feeder shall have an ampacity of not less than 125 percent of the full-load current of the highest rated motor of the group plus the sum of the full-load current ratings of the remaining motors in the group. The full-load current of the motor is taken from the *NEC*®, *Table 430.250*.

The motor with the largest full-load running current is the 25 hp motor. This motor has a full-load current rating of 68 amperes. The main feeder size for the highest rated motor, in compliance with *430.24*, is

$$68 \times 125\% = 85 \text{ amperes}$$

$$\text{Then } 85 + 54 + 28 = 167 \text{ amperes}$$

Table 310.15(B)(16) indicates that AWG 4/0 Type TW or Type THHN copper conductors can be used when using 60° terminations.

If all the terminations in the circuits are rated for 75°C, then the second column in *Table 310.15(B)(16)* can be used for all wire ampacities. *Article 110.14(C)(1)* states that motors with design letters B, C, or D may use the 75°C rating for terminations and wire ampacity. *Article 110.14(C)(1)(b)* states that circuits over 100 amperes or conductors larger than 1 AWG may use the 75°C rating. If all the other terminations in that circuit are rated at 75°C, then 2/0 THWN or THHN wire can be used.

Table C1 of *Annex C* shows that three 4/0 TW conductors can be installed in trade size 2 conduit. Three 4/0 THHN conductors can be installed in a trade size 2 conduit.

Annex C, Table C.1 lists that 3—2/0 THHN conductors can fit in a trade size 1-1/2 in.– (metric designator 41) EMT conduit. Many newly installed systems are rated for 75°C terminations, so this will enable the electrical installation to use smaller size wire and pipe than would be allowed with 60°C terminations.

Main Feeder Short-Circuit Protection

Article 430, Part IV, 430.62(A), states that a feeder that supplies motors shall be provided with overcurrent protection. The feeder overcurrent protection shall *not be greater* than the largest current rating of the branch-circuit protective device for any motor of the group, based on *Table 430.52*, plus the sum of the full-load currents of the other motors of the group.

The branch circuit for the 25 hp motor has the largest value of overcurrent protection. This value, as determined from *Table 430.52*, is a 119-ampere (68 × 1.75 or 125 ampere) fuse. The full-load current rating of the 20 hp motor is 54 amperes, and the full-load current rating of the 10 hp motor is 28 amperes. The size of the fuses to be installed in the main feeder circuit shall not be greater than the sum of 125 + 54 + 28 = 207 amperes.

Therefore, three 200-ampere, nontime-delay fuses are used for the feeder circuit. This procedure should be in conformance with *Example D8, Annex D* of the *Code*. Exceptions may be made if the fuses do not allow the motor to start or run.

Main Disconnecting Means

Article 430, Part IX, 430.109, lists several types of disconnecting means. The disconnecting means shall have a carrying capacity of at least 115% of the sum of the current ratings of the motors. See *430.110 (C1 and 2)*. Therefore, the 200-ampere fuses required as the overcurrent protection for the main feeder are installed in a 200-ampere safety switch.

    Wire types and sizes are selected by the ambient temperatures of the place of installation and the economics of the total installation, such as the minimum size of conduits, cost of the wire sizes, and labor cost for installing the various selections.

SUMMARY

Motor installation is one of the hardest calculations to perform and get all the components correct, in the proper location, and at the correct size. The *NEC®* guides you through the main components of the calculation, but you must know where to look and how to apply the proper codes. There are many facets to the correct installation, including feeder and feeder protection, branch-circuit and branch protection, conductor sizes and overcurrent protection, running overcurrent protection, and secondary circuit protection.

ACHIEVEMENT REVIEW

Refer to the following feeder circuit description for the calculations in items 1 through 5.

    A feeder circuit feeds three branch motor circuits. Branch motor circuit No. 1 has a load consisting of an induction motor with the following nameplate data:

No. 1

Squirrel-Cage Induction Motor	
230 Volts	15 Amperes
3 Phase	60 Hertz
Code Letter D	5 Horsepower
Temperature Rating 40° Celsius	

© Cengage Learning 2014

    Branch motor circuit No. 2 has a load consisting of an induction motor with the following nameplate data. (This motor is equipped with an autotransformer starting compensator.)

No. 2

Squirrel-Cage Induction Motor	
230 Volts	40 Amperes
3 Phase	60 Hertz
Code letter F	15 Horsepower
Temperature Rating 40° Celsius	

© Cengage Learning 2014

Branch motor circuit No. 3 has a load consisting of a wound-rotor induction motor with the following nameplate data.

Wound-Rotor Induction Motor	
230 Volts	22 Stator Amperes
3 Phase	26 Rotor Amperes
7 1/2 Horsepower	60 Hertz
Continuous Duty	Temperature Rating 40° Celsius

No. 3

© Cengage Learning 2014

1. Refer to Figure 23–4. Insert the answers on the diagram.
    a. Determine the running overload protection in amperes required for the motor in branch circuit No. 1.
    b. Determine the appropriate wire size (75° wire).

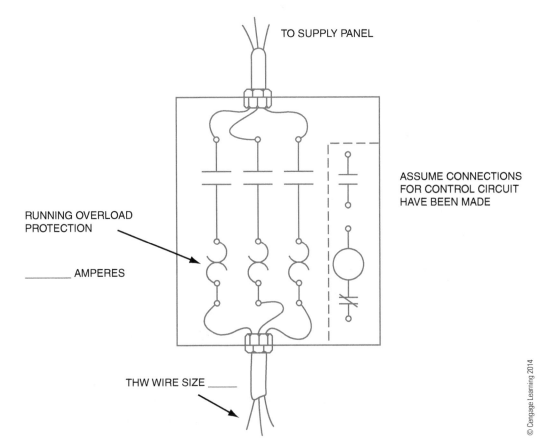

TO SUPPLY PANEL

ASSUME CONNECTIONS
FOR CONTROL CIRCUIT
HAVE BEEN MADE

RUNNING OVERLOAD
PROTECTION

_____ AMPERES

THW WIRE SIZE _____

FIGURE 23–4 Magnetic ATL motor starter switch for question 1.

© Cengage Learning 2014

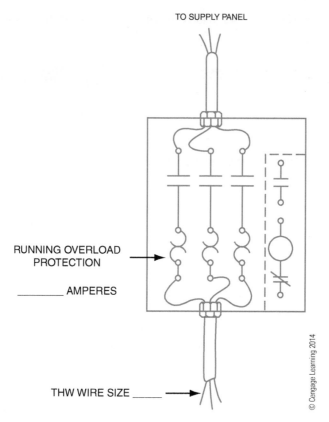

TO SUPPLY PANEL

RUNNING OVERLOAD
PROTECTION

_____ AMPERES

THW WIRE SIZE _____

© Cengage Learning 2014

FIGURE 23–5 Manual autotransformer starting compensator for question 2.

2. Refer to Figure 23–5. Insert the answers on the diagram.
   a. Determine the running overload protection in amperes required for the motor in branch circuit No. 2.
   b. Determine the appropriate wire size of the copper THW conductors. Note that the 15 hp squirrel-cage induction motor in this circuit is started by means of a starting compensator.

3. Refer to Figure 23–6. Insert the answers of (a) and (b) on the diagram.
   a. Determine the running overload protection in amperes required for the motor in branch circuit No. 3.
   b. Determine the appropriate wire size of the copper conductors.
   c. Determine the size of the conductors required for the secondary circuit of the wound-rotor induction motor in branch circuit No. 3. The secondary or rotor circuit feeds between the slip rings of the wound rotor and the speed controller. Indicate the size of the conduit. Use THW conductors.

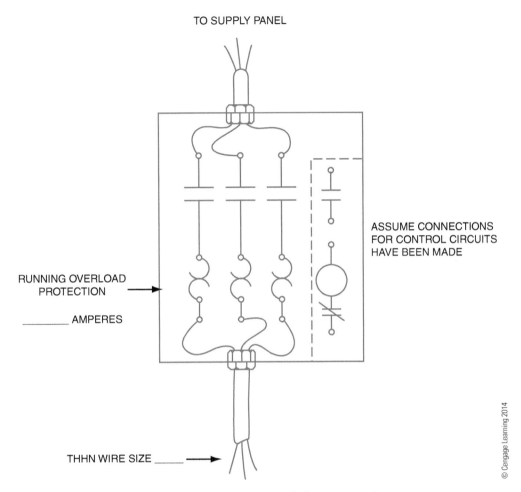

TO SUPPLY PANEL

ASSUME CONNECTIONS
FOR CONTROL CIRCUITS
HAVE BEEN MADE

RUNNING OVERLOAD
PROTECTION

_____ AMPERES

THHN WIRE SIZE _____

FIGURE 23–6 Magnetic ATL motor starter switch for question 3.

© Cengage Learning 2014

4. Refer to Figure 23–7. Insert the answers on the diagram.
   a. Determine the current rating in amperes of the fuses (nontime-delay) used as over-
      load protection for the main feeder circuit shown in the diagram.
   b. Determine the THHN conductor size for the main feeder switch.

5. Refer to Figure 23–8. Insert the answers on the diagram.
   a. Using THHN-type copper conductors, determine the size of the conductors and
      conduit required for the main feeder circuit that feeds the three branch motor cir-
      cuits. Indicate the sizes on the diagram.

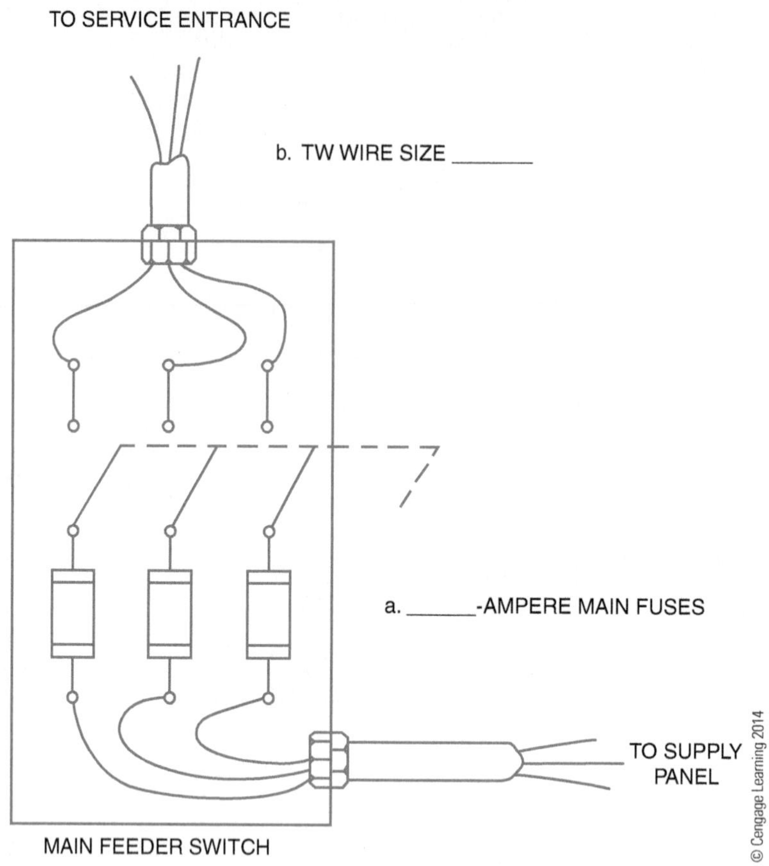

TO SERVICE ENTRANCE

b. TW WIRE SIZE _____

a. _____-AMPERE MAIN FUSES

TO SUPPLY
PANEL

© Cengage Learning 2014

MAIN FEEDER SWITCH

FIGURE 23–7 Fused disconnect for question 4.

b. Determine the size of fuses in amperes required for the starting overload protection for each of the branch circuits.

Motor Circuit No. 1 _____

Motor Circuit No. 2 _____

Motor Circuit No. 3 _____

c. Using THHN-type copper conductors, determine the size of rigid conduit required for each of the three branch circuits.

Motor Circuit No. 1 _____

Motor Circuit No. 2 _____

Motor Circuit No. 3 _____

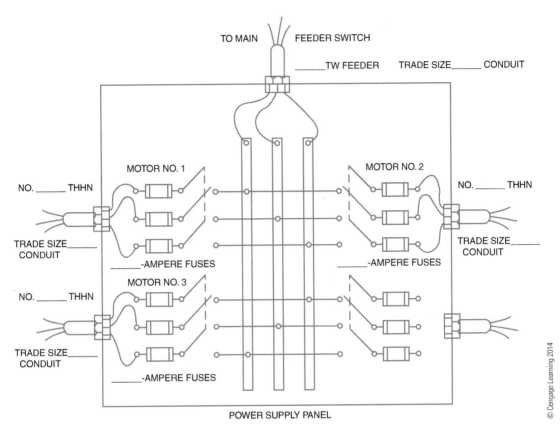

FIGURE 23–8 Feeder panel for question 5.

© Cengage Learning 2014

UNIT

24

SINGLE-PHASE MOTOR AND DC MOTOR INSTALLATION AND THE *NATIONAL ELECTRICAL CODE*®

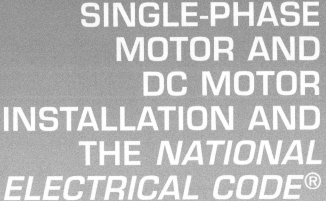

BJECTIVES

After studying this unit, the student should be able to

- determine the requirements for installation of AC single-phase motors.

- follow the *National Electrical Code*® requirements and find code references.

- install typical residential applications of single-phase motors to meet the code.

- follow *NEC*® installation requirements for DC motors.

NATIONAL ELECTRICAL CODE® *REGULATIONS*

NEC® 430.32(D)(2) states that any motor of 1 hp or less that is not permanently installed, is manually started, and is within sight of the starter location shall be considered as protected against overload by the overcurrent device protecting the conductors of the branch circuit. This branch overcurrent device shall not be larger than that specified in *Article 430, Part IV (Motor Branch Circuit, Short-Circuit and Ground-Fault Protection)*. An exception is that any such motor may be used at 120 volts or less on a branch circuit protected at not over 20 amperes. A distance of more than 50 ft is considered to be out of sight from the starter location. *NEC®* 430.32(B) covers motors of 1 hp or less that are automatically started. *NEC®* 430.32(B) (1) states that any motor of 1 hp or less that is started automatically shall have a separate overcurrent device that is responsive to the motor current. This overload unit shall be set to trip at not more than 125% of the nameplate full-load current rating for motors marked to have a temperature rise 40°C (104°F) or less, or with a service factor 1.15 or higher and not more than 115% for all other types of motors. Other options are to use motors with integral thermal protection or impedance-protected motors.

Single-phase motor installation must follow the *National Electrical Code®*. As you have seen throughout the book, there are many different types of motors and these can be installed in residential applications as well as non-dwelling units. *Article 430* governs the installation of motors in all applications. The motors include AC single-phase motors as well as three-phase motors and DC motors. Single-phase AC motors are most common in residential applications and include most household applications such as garbage disposal motors, furnace blower motors and air-conditioning compressor motors.

Residential Single Phase AC Motors

To get a good view of the motor installation requirements, refer to *Figure 430.1* in the *NEC®*. See Figure 24–1. For small residential uses, the motor circuits start at the branch circuit overcurrent protective device (OCP). A branch circuit is generally defined as the circuit conductors that extend from the last circuit protective device for overcurrent, to the power utilization (outlet). Here we start with the motor and work back to the branch-circuit protection.

As the motor is installed, the nameplate information is needed. As noted in *Article 430.7*, all of the information is needed to properly install the motor in a safe and efficient matter. A few of the items are of particular importance. *Item 13* in *Article 430.7(A)* indicates that some motors may have a thermal protector installed on the motor. If there is an internal (integral) protector, it shall be marked "thermally protected." If such a motor is less than 100 watts and complies with 420.32(B)(2) (if the protector is connected to a control circuit rather than the power circuit), they may be marked "TP" *Item 14* states that a motor may be impedance protected—meaning that the internal impedance of the motor windings is high enough to protect the motor from overheating. If these motors are 100 watts or less and they comply with 430.32(B)(4), then they

General, 430.1 through 430.18 — Part I
Motor Circuit Conductors, 430.21 through 430.29 — Part II
Motor and Branch-Circuit Overload Protection, 430.31 through 430.44 — Part III
Motor Branch-Circuit Short-Circuit and Ground-Fault Protection, 430.51 through 430.58 — Part IV
Motor Feeder Short-Circuit and Ground-Fault Protection, 430.61 through 430.63 — Part V
Motor Control Circuits, 430.71 through 430.74 — Part VI
Motor Controllers, 430.81 through 430.90 — Part VII
Motor Control Centers, 430.92 through 430.98 — Part VIII
Disconnecting Means, 430.101 through 430.113 — Part IX
Adjustable Speed Drive Systems, 430.120 through 430.128 — Part X
Over 600 Volts, Nominal, 430.221 through 430.227 — Part XI
Protection of Live Parts—All Voltages, 430.231 through 430.233 — Part XII
Grounding—All Voltages, 430.241 through 430.245 — Part XIII
Tables, Tables 430.247 through 430.251(B) — Part XIV

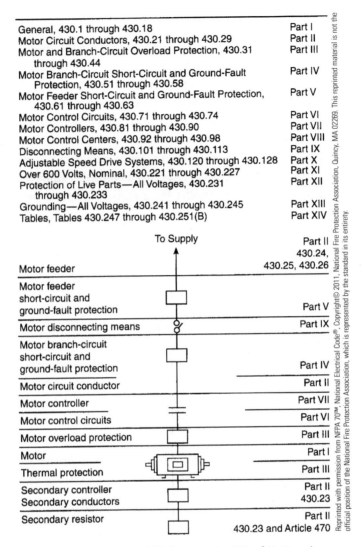

To Supply

Motor feeder — Part II, 430.24, 430.25, 430.26

Motor feeder short-circuit and ground-fault protection — Part V

Motor disconnecting means — Part IX

Motor branch-circuit short-circuit and ground-fault protection — Part IV

Motor circuit conductor — Part II

Motor controller — Part VII

Motor control circuits — Part VI

Motor overload protection — Part III

Motor — Part I

Thermal protection — Part III

Secondary controller — Part II
Secondary conductors — 430.23

Secondary resistor — Part II, 430.23 and Article 470

FIGURE 24-1 Figure 430–1 in article 430 of National Electrical Code®.

may be marked "ZP." As indicated in *430.32(B)(4)*, these motors have sufficient impedance to prevent overheating even if the motor does not start and stays connected to the branch circuit. Typical motors of this type are clock motors less than 1/20 hp. *NEC® 430.32(D)(2)(a)* states that for 1 hp or less, and not automatically started these "ZP" protected motors can be considered protected by the branch-circuit protection, and the *Exception* states that any such motor can be protected by a 120-volt branch circuit not over 20 amps. Therefore, most small clock motors need no further protection as long as they are not permanently installed. Other larger motors are protected as follows.

Garbage Disposal Motors

The branch-circuit protection also provides the motor branch circuit, short circuit, and ground-fault protection" as required in the *NEC®* under *Part IV of Article 430*. The motor circuit conductor as required in *Part II of Article 430* is the wiring to the motor itself. This can include the cord connected with a plug to a receptacle under the sink (one method of installation). Other installations may have a permanent connection with flexible conduit. Motor controllers are covered in *Part VI of Article 430*. *NEC®* 430.81(A) describes small motors that are stationary and 1/8 hp or less, such as a clock motor, stating the branch-circuit overcurrent device can serve as the motor controller. *Article 430.82* states that a controller must be capable of starting and stopping the motor. The motor controller for a garbage disposal motor installation is a switch, usually found on the backsplash, next to the sink. It is capable of starting and stopping the motor. *Article 430.83* outlines the rating of the controller. *NEC®* 430.83(A) states that the controller must have a horsepower rating unless the motors fall under *Part B, C, or D. Part C* refers to the small stationary motors that are 2 hp or less and 300 volts or less. A general-use snap switch may be used as a controller if it has an ampere rating at least twice the full-load current rating of the motor. For AC motors, the general-use snap switch is rated only for AC, not AC/DC snap switches, where the motor full-load rating is not more than 80% of the switch rating. The controller is not required to open *all* conductors to the motor but in this case would open the hot conductor. These single-phase motors usually have built-in integral overload protection in the form of a current sensor that has a resettable breaker. In case the food gets jammed in the mechanism, the motor is protected. The overload protection required by the *NEC®* to prevent overheating of the motor and its wiring is covered in *Article 430, Part III*. In the case of a garbage disposal motor, most are referred to in *Article 430.32* either *Part A* for motors more than one horsepower or *Part B* for motors 1 hp or less. Either part refers to *Section 2*, which states that a thermal protector in the motor must protect it from dangerous overheating due to overloads or failure to start.

Furnace Motors

Small residential furnace installations fall under the same information as the garbage disposal motors as related to *NEC® Article 430, Parts I, II, III,* and *IV*. Forced air gas or oil burner furnace installations typically have one or more motors that provide combustion air and also a fan that circulates the warm air to the inhabited space. Electricians typically connect the branch-circuit power and the local disconnect (a switch) for the furnace power, and all the rest of the controls for the fan are internal to the furnace controls.

Air Conditioning and Refrigeration Equipment

Article 440 of the *NEC®* is pertinent to cooling systems. *Article 440.2* states that equipment that has a running overcurrent protection device that allows a continuous current greater than the normal full-load current, shall be marked with the higher current and referred to as the branch-circuit selection current. This branch-circuit selection current is the current to be used in sizing branch-circuit conductors, disconnects, controllers, branch-circuit, short-circuit and ground-fault protection. This is also noted in *440.6(B)*. If the equipment does not have a nameplate, then the nameplate current on the hermetic refrigerant motor compressor shall be used for calculation as noted in *440.6(A)*. A disconnecting means is required outside next to a condenser unit that contains a fan and a compressor motor. The disconnect is rated according to *Article 440.12*—at least 115% of the appropriate current. Branch-circuit overcurrent protection is sized according to *Part III of Article 440*. Generally, a rating not exceeding 175% of the rated load current is used for protecting against short circuits and ground faults. Branch-circuit conductors are sized to 125% of the load current according to *Part IV of Article 440*. Motor controllers are sized according to *Part V of Article 440*. Generally, the controllers must be rated to carry the full-load current plus the locked rotor current of the motor. Motor compressor overload must follow *Part VI of Article 440*. Overload relays must trip at no more than 140% of rated current. A thermal protector integral with the motor is pre-rated to the motor. A fuse or inverse time circuit breaker is rated at 125% of full current as protection.

Other Single-phase Motor Installations

In other motor installations in residential or nonresidential applications, the *NEC*® is the guide to proper protection for the motor, to prevent overheating and causing a fire. This protection is covered in *Article 430, Part III*. The wiring to the motor is also protected according to *Article 430, Part IV*, of the *NEC*® under the heading *Motor Branch Circuit Short Circuit and Ground Fault Protection*. In general, the requirements for protections are detailed in *430.52(A)* and *(B)*. The branch-circuit protection is based on *NEC*® *Table 430.52* as a percentage of full-load current. Remember that the full-load current referred to here is the *Code* book value as noted in *430.6*. The full-load current is listed in *Table 430.248* for single-phase motors. If using a typical inverse time breaker, the percentage for single-phase motors is 250% of the *Code* book value for the size of the motor at the connected voltage. This high percentage is allowed so that the breaker does not trip every time the motor starts, because the motor draws many times the rated current as it starts. The high percentage is considered proper to protect the wiring from short circuit, and faults to ground in the branch circuit conductors.

DC Motors

DC motor installations also fall under *Article 430* of the *NEC*®. As with all these motor installations, refer to *Article 430.6* to determine which current value to use for the calculations needed. For DC motors, the values given in *Table 430.247* are used for calculations except for the separate motor overload (running overcurrent protection) calculations that do use the actual nameplate data to protect the specific motor installed. Most DC motor installations recently installed have electronic controls in which much of the overload protection is part of the solid-state control circuitry. As indicated in *Article 430.22(A)* for single DC motors supplied from a DC rectifier power source, the ampacity to the input of the solid-state controller must not be less than 125% of the rated input current of the controller. If the field control for the DC motor is fed by a half-wave rectifier, the field wiring must be at least 190% of the motor full-load current. If the field is supplied by a full wave rectifier configuration, the field wiring must be not less that 150% of the nameplate data.

Refer to *Table 430.37* to determine where running protection is required for DC motors installed under various wiring system supplies, and how many overload units are required.

Summary

Single-phase motors are often used in small motor needs for typically light loads, sometimes up to 10 hp. These installations are very common in small to moderate residential applications where single-phase power supplies are used. Where these motors are installed, the *NEC*® is followed to prevent dangerous overheating of motors and branch-circuit faults that could ignite fires. *Figure 430.1* 0f the *NEC*® details, where each item for protection is located within *Article 430*. *Table 430.5* lists other articles that have specific requirements for motor installation with special equipment or special occupancies. DC motors are less commonly installed, but the provisions of *Article 430* apply to these motor installations as well.

Achievement Review

1. Describe the difference between motor short-circuit protection and motor overload protection. _____

2. When determining the short-circuit protection percentage for a 1 hp single-phase motor that has a name plate current of 14.7 amps, what value of current would you use?

   _____

3. If a single-phase motor supplied by a two-wire, 120-volt circuit with one conductor grounded, needed an overload unit, in which conductor would the unit be installed?

   _____

4. An impedance-protected motor would be protected against overcurrent by the

   _____ .

5. Where installing a single-phase motor that is not Z protected, the short-circuit ground-fault protection is rated at _____ % of _____ value when using an inverse time breaker.

6. If you have to install a constant voltage DC motor using an inverse time breaker for short-circuit protection, then use a breaker rated at _____ % of *Code* book value.

7. The branch-circuit conductor that supplied power to a single-phase cap start motor should be rated at _____ % of the _____ _____ value for the motor at connected voltage and rated hp.

8. The standard controller for a motor used in residential garbage disposal is a _____.

9. A single-phase air-conditioner outdoor condenser/compressor would have a _____ connected within sight of the unit.

10. If a continuous duty motor more than 1 H.P. has a thermal protector protecting it from overload and the service factor is 1.15, then the trip value of the OL would be _____% of the nameplate current of the motor.

UNIT 25

MOTOR MAINTENANCE

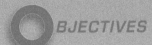

BJECTIVES

After studying this unit, the student should be able to

- perform routine inspection and maintenance checks of motors.
- perform the following simple tests:
  - measure insulation resistance.
  - use a growler to locate short-circuited coils.
  - perform continuity checks for open-circuited coils.
  - measure balance to determine phase currents under load.
  - measure speed variation.
- replace and lubricate sleeve and ball bearings according to manufacturers' directions.
- lubricate motors according to manufacturers' directions.

PREVENTIVE MAINTENANCE

Most electrical equipment requires planned inspection and maintenance to keep it in proper working condition. Periodic inspections prevent serious damage to machinery by locating potential trouble areas. Observant personnel fully use their senses to diagnose and locate problems in electrical machinery: *smell* directs attention to burning insulation; *touch* detects excessive heating in windings or bearings; *hearing* detects excessive speed or vibration; and *sight* detects excessive sparking and many mechanical faults.

Sensory impressions usually must be supplemented by various testing procedures to localize the trouble. A thorough understanding of electrical principles and the efficient use of test equipment are important to the electrician in this phase of troubleshooting.

PERIODIC INSPECTIONS

The ideal motor maintenance program aims at preventing breakdowns rather than repairing them. Systematic and periodic inspections of motors are necessary to ensure optimal operating results. In a good preventive maintenance program with detailed checks, the person in charge should have a history on file for every motor in the plant. Entries should include inspection dates, descriptions of repairs, and the costs involved. When the record indicates that a motor has undergone excessive and/or costly repairs, the causes can be determined and corrected.

Inspection records also serve as a guide to indicate when motors should be replaced because of their high operational cost. They also reveal faulty operating conditions, such as misapplication or poor drive engineering.

Inspection and servicing should be systematic. However, the frequency of inspections and the degree of thoroughness may vary, as determined by the plant maintenance engineer. Such determinations are based on (1) the importance of the motor in the production scheme (if the motor fails, will production be slowed seriously or stopped?), (2) the percentage of the day the motor operates, (3) the nature of the service, and (4) the motor's environment. An inspection schedule therefore must be flexible and adapted to the needs of each plant. Equipment manufacturers' specifications and procedures should be consulted and followed.

The following schedule, which covers both AC and DC motors, is based on average conditions insofar as operational use and cleanliness are concerned. (Where dust and dirty conditions are extremely severe, open motors may require a certain amount of cleaning every day.)

Every Week

1. Examine commutator and brushes, AC and DC.
2. Check oil level in bearings.

3. See that oil rings turn with shaft.

4. See that exposed shaft is free of oil and grease from bearings.

5. Examine the starter switch, fuses, and other controls; tighten loose connections.

6. See that the motor is brought up to speed in normal time.

Every Six Months

1. Clean motor thoroughly, blowing out dirt from windings, and wipe commutator and brushes.

2. Inspect commutator clamping ring.

3. Check brushes and replace any that are more than half worn.

4. Examine brush holders, and clean them if dirty. Make certain that brushes ride free in the holders.

5. Check brush pressure.

6. Check brush position.

7. Drain, wash out, and replace oil in sleeve bearings.

8. Check grease in ball or roller bearings.

9. Check operating speed or speeds.

10. See that end play of shaft is normal.

11. Inspect and tighten connections on motor and control.

12. Check amperage input and compare it with normal.

13. Examine drive critically for smooth running, absence of vibration, and worn gears, chains, or belts.

14. Check motor foot bolts; end-shield bolts; pulley, coupling, gear, and journal set-screws; and keys.

15. See that all covers and belt and gear guards are in place, in good order, and securely fastened.

Once a Year

1. Clean out and renew grease in ball or roller bearing housings.

2. Test insulation by megohmmeter.

3. Check air gap of rotor.

4. Clean out magnetic dirt that may be clinging to poles.

5. Check clearance between shaft and journal boxes of sleeve-bearing motors to prevent operation with worn bearings.

6. Clean out undercut slots in commutator. Check the commutator for smoothness.

7. Examine connections of commutator and armature coils.

8. Inspect armature bands.

MEASUREMENT OF INSULATION RESISTANCE

The condition of electrical insulation is an important factor in the maintenance of motors and alternators. Moisture, dirt, chemical fumes, and iron particles all cause deterioration of the insulation used on the windings of stators and rotors. Motors operating under adverse conditions require periodic tests to ensure continuous and satisfactory operation. Although severe conditions can be detected by touch, sight, or smell, often it is necessary to use more accurate measures of the condition of the insulation at any given time.

The value of insulation resistance in megohms (1,000,000 ohms) is used as an indication of insulation efficiency. Successive readings taken and recorded under the same test conditions document the insulation history of a unit and serve as an index of insulation deterioration. A *megohmmeter,* commonly called a *Megger®,* shown in Figure 25–1(A) and (B), is used to measure insulation resistance. In general, the instrument develops a voltage that is applied to the insulation path. The amount of current in this path is shown on a sensitive *microammeter* calibrated in megohms. When the instrument leads are not connected, the microammeter reading should be infinity. Specific operating instructions are provided with the instrument by the manufacturer.

The megohmmeter ground lead is connected to the frame of the machine. The ungrounded lead is connected to any metallic part of the winding being tested, such as any terminal of the coil circuit. No other external paths must exist in parallel circuits. For this reason, the windings being tested should be isolated by disconnecting them from other parts of the circuit. The megohmmeter reading, then, is the resistance of the insulation between the number of coils in the circuit and the frame of the machine.

Major motor manufacturers apply insulation resistance for 660 volts for low-voltage motors below 600 volts. A method of determining the normal insulation resistance value is given in the following formula. This nonexact formula has been derived from experience. It is believed that the Institute of Electrical and Electronic Engineers, Inc. (IEEE) accepted this general theory from a manufacturer of megohmmeters.

Caution: Disconnect electronic devices attached to units under test.

$$Rm = KV + 1$$

$$Rm = \text{Recommended megohms } 40°C$$

$$KV = \text{Rated terminal voltage in kilovots}$$

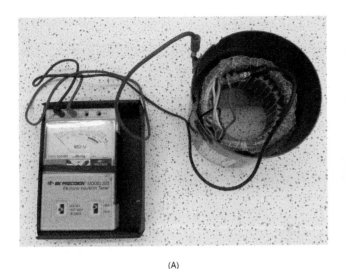

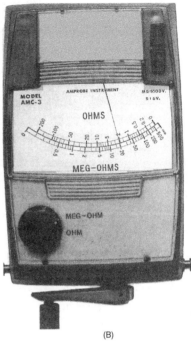

FIGURE 25–1 (A) Connections for testing motor winding resistance to the motor frame. (B) Megohmmeter used with hand crank operation.

This method is usually stated as 1 megohm per 1000 volts plus 1 megohm for a 60-second test at 40°C (104°F).

A very low value of insulation resistance indicates defective insulation. The electrician should begin an immediate check to localize the defective insulation. This is done by disconnecting various coils from a series, parallel, wye, or delta combination, and repeating the insulation resistance test on each isolated coil. A very low value of resistance may indicate a grounded coil (complete breakdown of insulation at some point).

A value of insulation resistance slightly below the recommended approximate value does not necessarily indicate that immediate repair is required. The electrician should take a series of readings at weekly intervals to detect any progressive decrease or sudden drop in the insulation resistance. If the resistance continues to decrease, then the fault is to be located without delay. A slightly low but constant value of resistance should not cause concern.

Note: If a motor has been reading 10 megohms on periodic tests and suddenly drops to 0.2 megohms the motor may be unsafe to start. New motors read around infinity (the highest resistance on the meter scale), or slightly lower, but during use, this resistance lowers to a more steady reading.

TESTING FOR SHORT-CIRCUITED COILS

Open-circuited coils on rotors or stators can be located by continuity (one end to the other) tests. Short-circuited coils are located easily by the use of a growler. A *growler* is an instrument consisting of an electromagnetic yoke and winding excited from an AC source. The yoke is placed across a section of slots containing the winding being tested. The yoke winding acts as the primary winding of a transformer, and the winding being tested acts as the secondary (Figure 25–2).

If a turn or coil is short-circuited, the resulting current rise in the primary (yoke) circuit is indicated on the ammeter. If the current is permitted to exist for a short period of time, the defective turn or coil can be identified by the heat developed at the defective point. The stator windings of both alternators and motors can be tested by this method.

The field coils of alternators can be tested using the voltage-drop method described for DC machinery testing. With a given value of current in the field circuit, the voltage drops on individual field coils should be approximately equal. If a voltage-drop difference is in excess of 5 %, the coil should be investigated for shorted turns. The presence of full line voltage across a single coil indicates an open circuit in that coil.

The field coils of alternators can be checked for impedance by applying a high-frequency low voltage to each coil and measuring the current. The currents in the coils should be equal. The presence of a high current usually means that there are shorted turns somewhere in the coil.

BALANCE TEST

The current in the individual phases of three-phase motors must be equal. Figure 25–3 shows the connections necessary to make a simple balance test that measures the phase currents under load. This test may be made in the electric shop before a motor is installed.

The fused three-pole switch at the left of the diagram is used to start the motor. The second three-pole switch removes the ammeter from the circuit during the starting period when the current input is very high. This switch is closed before the motor is started. When the motor reaches its rated speed, the ammeter switch is opened. The current in each phase, therefore, is indicated on the ammeter. For a motor operating normally, the three line currents are equal. A high reading in one phase may indicate shorted turns. If one phase shows no current, then the motor is operating as a single-phase motor. High but equal current readings in all three phases indicate an overloaded motor.

Figure 25–4 shows the use of a more convenient method of making a balance test. A clamp-on ammeter is used to take readings in each phase of a motor in actual operation under normal load.

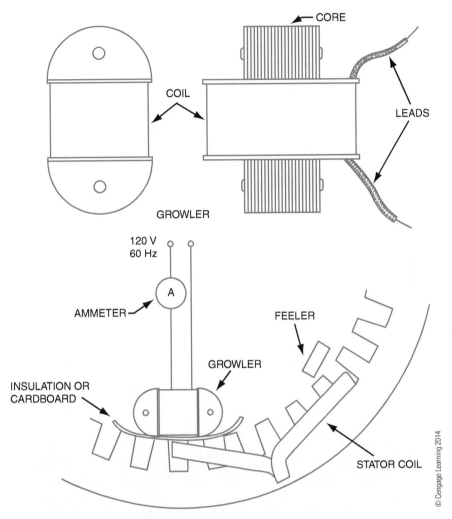

FIGURE 25–2 A growler for testing shorted coils.

MEASUREMENT OF SPEED VARIATIONS

Deviations from the rated speed of a motor under load are an indication of improper mechanical loading or faulty conditions within the motor. A *tachometer* (Figures 25–5 and 25–6) is an instrument used to check the speed of a motor. Other types of instruments are also used. The tachometer in Figure 25–6 is held by hand to the end of the motor shaft to measure shaft speed.

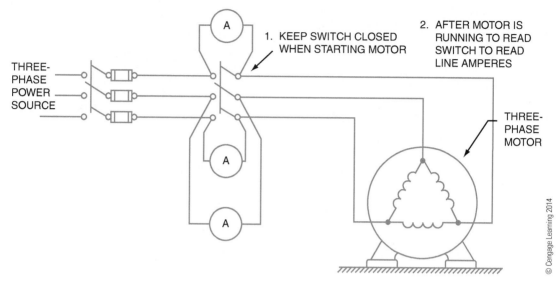

FIGURE 25-3 A balanced current motor test.

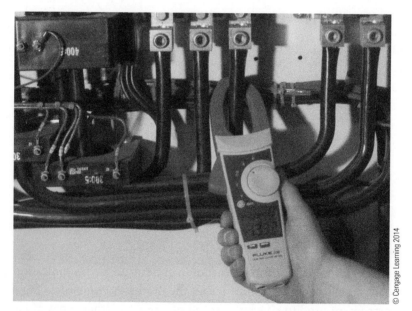

FIGURE 24-4 Use a clamp-on ammeter to test motor power system.

© Cengage Learning 2014

FIGURE 25-5 A strobe tachometer and a photoreflective tachometer.

SQUIRREL-CAGE ROTORS

The bars of a squirrel-cage rotor can be broken if the motor is subjected to severe jarring, vibration, or overheating. Bars that are dislodged must be put back in place securely so that movement is impossible. The presence of broken bars can be detected by the use of a growler. (The rotor must be removed to make this test.)

A test can be performed that does not require the removal of the rotor to detect broken or open rotor bars. This test consists of exciting one phase of the stator winding with 25% of its normal voltage. Enough voltage is used to give a suitable indication on an ammeter in series with the winding. By turning the rotor slowly by hand, any variations in stator current can be observed on the ammeter. Any current variation in excess of 3% usually indicates open bars in the rotor (Figure 25-7).

BEARINGS

The type of bearings used in a motor depends on the cost of the motor and the characteristics of the load. Sleeve and ball bearings are used in both AC and DC motors. Excessive wear on the bearings reduces the concentricity of the stator and rotor sections. In small motors, with the power off, excessive wear can be detected by manually attempting to move the shaft of the rotating member in a lateral direction. The amount of play in the shaft is an indication of bearing

FIGURE 25–6 Mechanical tachometer used to measure shaft RPM.

© Cengage Learning 2014

wear. For large motors, bearing wear and the resulting deviation in concentricity of the stator and rotor can be detected by measuring the air gaps over several points around the periphery of the rotor. Severe bearing wear on both large and small motors may result in actual contact between the rotor and the stator.

The motor must be disassembled to repair bearings. This type of repair requires special tools, and the job should not be attempted without them.

A sleeve bearing is removed by dismantling the baffles inside the end shield, removing the oil well cover plates, and removing the oil ring clips. The bearing lining is then tapped out using a short length of pipe stock or a special split fitting that locks inside the oil ring slot.

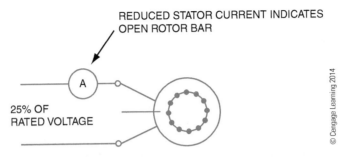

REDUCED STATOR CURRENT INDICATES
OPEN ROTOR BAR

A

25% OF
RATED VOLTAGE

© Cengage Learning 2014

FIGURE 25-7 An ammeter used to test for open rotor bars in a squirrel-cage rotor.

Ball Bearings

Ball bearings are pressed fit to the motor shaft and usually should be removed from the shaft only when it is necessary to replace the bearing. To inspect the ball bearings, the end bells are removed and the rotor, the rotor shaft, and the bearing assembly are taken from the stator. In some motors, the bearing housings have removable bearing caps so that it is possible to remove the bearing without removing the end bells.

If a ball bearing must be replaced, a bearing puller is normally used to prevent damage to the shaft (see Figure 25–8). The electrician must be very careful when placing a new bearing on the shaft so that neither the bearing nor the shaft is damaged (Figure 25–9). A ball bearing race

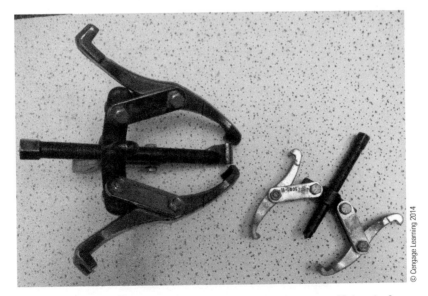

© Cengage Learning 2014

FIGURE 25-8 Bearing pullers. Use jaws to pull bearing off the shaft.

WRONG:

BEARING SHOULD NOT BE FORCED ON SHAFT BY TAPPING ON THE OUTER RINGS. IT SHOULD NOT BE FORCED ON A BADLY WORN SHAFT OR ON A SHAFT THAT IS TOO LARGE.

RIGHT:

BEARING IS PROPER SIZE FOR SHAFT AND IS BEING TAPPED LIGHTLY INTO PLACE BY MEANS OF A METAL TUBE THAT FITS AGAINST THE INNER RING. DO NOT POUND ON THE BEARING.

© Cengage Learning 2014

FIGURE 25–9 Bearing installation.

must be placed on the shaft so that the race is exactly square with the shaft. The shaft must be in perfect condition because even the slightest burr will cause trouble. *Pressure must not be applied to the outer race of the bearing.* Pressure applied to install the bearing must be applied evenly on the diameter of the inner race (Figure 25–10). Light tapping is recommended. A piece of pipe stock slightly greater than the shaft diameter is used to press on the new bearing. The bearing can be warmed to a temperature of 150°F (66°F) to simplify the process.

Courtesy of General Bearing Corporation®

FIGURE 25–10 Single-row, snap-ring ball bearing.

New ball bearings must not be cleaned prior to installation. Dust or dirt must not enter the bearing during the installation.

LUBRICATION

Several methods of lubricating motors are used. Small motors with sleeve bearings have oil holes with spring covers. These motors should be oiled periodically with a good grade of mineral oil. Oil with a viscosity of 200 seconds Saybolt (approximately SAE 10) is recommended.

The bearings of larger motors are often provided with an oil ring that fits loosely in a slot in the bearing. The oil ring picks up oil from a reservoir located directly under the ring. Under normal operating conditions, the oil should be replaced in the motor at least once a year. More frequent oil replacement may be necessary in motors operating under adverse conditions. In all cases, avoid excessive oiling; insufficient oil can ruin a bearing, but excessive oil can cause deterioration of the insulation of a winding.

Many motors are lubricated with grease. Periodic replacement of the grease is recommended. In general, the grease should be replaced whenever a general overhaul is indicated, or sooner if the motor is operated under severe operating conditions.

Grease can be removed by using a light mineral oil heated to 165°F (74°F) or a solvent. Any grease-removing solvents should be used in a well-ventilated work area.

Ball-Bearing Lubrication

The following table indicates recommended intervals for regreasing of ball-bearing units. It is important that the correct amount and type of grease be used in ball bearings. Too much grease can cause overheating.

	Horsepower		
Service	1/4–7 1/2	10–40	50–150
Easy	7 yrs.	5 yrs.	3 yrs.
Standard	5 yrs.	2 yrs.	1 yr.
Severe	3 yrs.	1 yr.	6 mos.
Very severe	6 mos.	3 mos.	3 mos.

© Cengage Learning 2014

Fractional horsepower motors often contain sealed bearings. With normal service, these bearings do not require regreasing. When regreasing is required for unshielded bearings, the manufacturer's specifications and directions must be followed.

Some motors with ball bearings are provided with pressure fittings, and a grease gun is used to lubricate the bearings. Remove the bottom plug when doing this. Because of the wide variation in the design of industrial motors, the electrician should consult the comprehensive lubrication manuals published by electrical machinery manufacturers to ensure the proper lubrication of all types of motors.

BALL AND ROLLER BEARING MAINTENANCE

Cleaning Out Old Grease

It is best to remove the bearing, if possible. However, when cleaning a bearing in place, remove as much of the old grease as possible, using a rag or brush free from dirt or dust. Flush out the housing with clean, hot kerosene (110°F to 125°F [43°C to 52°C]); clean, new oil; or solvent.

After the grease has been removed, flush out the bearing with light mineral oil to prevent rust and to remove all traces of cleaning fluid. Allow to drain thoroughly before adding new grease. When the bearing is removed from the case, wash out the hardened, rancid grease from the housing and bearing as follows:

1. Put on safety glasses or a face shield for protection.
2. Soak the bearing in hot kerosene, then remove it.
3. Rotate the bearing slowly using an air hose. (Fast rotation may score the balls without lubrication.)
4. Dip the bearing in clean kerosene, light oil, or solvent.
5. Rotate the bearing slowly again using an air hose.
6. Wash the bearing again with clean kerosene or solvent.
7. Rotate the bearing in hand, and check for smoothness.
8. If the bearing is smooth, repack it with grease.
9. If the bearing is not smooth, but is in good condition, there is still some hardened grease in it. Repeat operations 6 and 7.

Lubricating Bearings with Pressure Relief Plug

1. Wipe fitting and plug clean.
2. Remove relief plug at bottom of bearing (Figure 25–11).
3. With shaft in motion (if possible), force grease into the top grease fitting, catching old grease in a pan. Add grease until new grease appears at the pan.
4. With a screwdriver, open the relief hole. Do not push the screwdriver into the gearing housing.

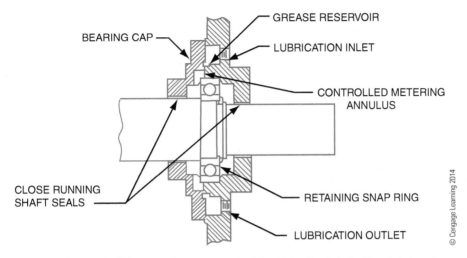

FIGURE 25-11 Ball bearing housing assembly. Note the lubrication inlet and outlet.

5. Allow bearing to run with relief plug out to remove pressure.

6. After the grease stops running out, and there is no longer any pressure in the bearing, replace the relief plug.

Lubricating Bearings without Relief Plug

1. Install a grease fitting with safety vents.

2. With the motor running, pump in grease slowly until a slight bleed shows around the seal or safety vent.

3. If it is necessary to lubricate while bearing is standing, fill the bearing with grease from one-fourth to one-half of its capacity.

Note: Be sure not to overgrease and to keep the grease clean.

Relubrication with Oil

Oil is always subject to gradual deterioration from use and contamination from dirt and moisture. Because of this, regular intervals for cleaning bearings must be maintained.

1. After draining the used oil, flush out the bearing. This can be done by using a new charge of the lubricant used for regular lubrication. Run the machine for 3 to 5 minutes, and drain again. Units used where there is sawdust or dirty conditions may require two flushings.

2. Fill with new oil to the proper level. Be sure new oil is from a clean container, and that no dirt is pushed into the filler plug while filling. Lubricant must be kept clean.

3. Check the seals to see that they are effective to prevent leakage or the entry of outside dirt to the bearing. It is important to keep the lubricant clean and the bearing flushed out so that it will be clean. Always read and follow the manufacturer's instructions whenever possible.

OILS FOR GEAR MOTORS

A gear motor is a self-contained drive made up of a ball-bearing motor and a speed-reducing gear box. It is designed to take advantage of the electrical efficiency of the high-speed motor and the transmission efficiency of gears.

The front motor bearings are generally grease lubricated and require the same attention as standard ball-bearing motors. The rear bearings, gear box bearings, and the gears themselves are almost always splash lubricated from the same oil supply reservoir in the lower section of the gear unit.

Oil seals at each bearing prevent oil leakage into the motor windings and out along the drive shaft. The precision-cut gears require carefully selected lubricating oils. Use only top-grade oils of the viscosity called for by the manufacturer of the gear motor.

SUMMARY

Proper motor maintenance extends the life of the motor. Preventive maintenance helps you spot potential problems and correct them before the motor breaks down and causes delays in production or failures at critical times. If a motor does fail, it may be possible to repair minor problems in the maintenance shop and to recondition the motor for use as a spare. Follow manufacturers' directions when lubricating the motor bearings to keep the motor running smoothly without excessive current draw or excessive heat. Check for proper ventilation, proper voltage, and proper current when performing maintenance checks on the motor.

ACHIEVEMENT REVIEW

A. Select the correct answer for items 1 through 10, and place the corresponding letter in the space provided.

1. Periodic inspection of motors, controls, and other electrical equipment is important because it _____
   a. gives advance notice of impending trouble.
   b. is required by the job standards.
   c. is a requirement of supervision.
   d. completes a day's work.

2. Careful motor troubleshooting requires the use of _____
   a. the sense of smell.
   b. the sense of touch.
   c. hearing and vision.
   d. all of the above

3. The most accurate method of testing insulation resistance uses a(n) _____
   a. megohmmeter.
   b. growler.
   c. ohmmeter.
   d. tachometer.

4. Insulation resistance is measured in _____
   a. megawatts.
   b. megohms.
   c. kilohms.
   d. kilovolts.

5. A very low value of insulation resistance indicates_____
   a. a good operating condition.
   b. a fair operating condition.
   c. an immediate investigation is necessary.
   d. that the measuring instrument is at the wrong setting.

6. Short-circuited coils are located efficiently by the use of a(n) _____
   a. megohmmeter.
   b. growler.
   c. ohmmeter.
   d. clamp-on ammeter.

7. On a balance test for phase currents where one phase shows a higher reading, the
   probable cause is _____
   a. open turns.
   b. shorted turns.
   c. worn bearings.
   d. the need for rotor balancing.

8. If a three-phase AC induction motor is operating with current in only two line leads, then _____
   a. the motor is operating as a two-phase motor.
   b. the insulation is overloaded.
   c. there is an open circuit in the stator.
   d. there is an open circuit in the rotor.

9. A properly sized ball-bearing race is mounted on a shaft correctly by tapping _____
   a. a metal tube on the outer ring.
   b. the inner ring with a hammer.
   c. the outer ring with a hammer.
   d. a metal tube on the inner ring.

10. A ball-bearing, 10 hp, three-phase, 230-volt induction motor operating under very severe conditions should be greased about every _____
    a. 3 months.
    b. 6 months.
    c. 12 months.
    d. month.

B. Complete each of the following statements.

11. The instrument used to measure insulation resistance is operated with the test leads unconnected. The meter reading should be _____.

12. The measurement of individual phase currents in the operation of a three-phase induction motor is called a(n) _____.

13. The rotor and stator concentricity in three-phase motors and alternators can be determined by measurements of the _____.

14. Ball bearings should be removed from a motor shaft using a bearing _____.

15. Pressure should never be applied to the _____ race of a ball bearing.

16. Ball bearings are lubricated with _____.

17. Insufficient oil can ruin a bearing, but excessive oiling can ruin the _____.

UNIT

SUMMARY REVIEW
OF UNITS 19–25

26

OBJECTIVE

- To provide the student with an opportunity to evaluate the knowledge and understanding acquired in the study of the previous seven units.

1. List four applications of selsyn units.

   a. _____

   b. _____

   c. _____

   d. _____

2. Draw a schematic wiring diagram of a selsyn system consisting of a transmitter and one selsyn receiver unit.

3. Explain the operation of a simple selsyn system consisting of one transmitter and one selsyn receiver unit. _____

   _____

   _____

_____

_____

_____

4. Explain the purpose of a differential selsyn unit.

_____

_____

_____

_____

_____

5. Draw a schematic wiring diagram of a selsyn system with one selsyn exciter unit, a differential selsyn unit, and one selsyn receiver unit.

6. Insert the correct word or phrase to complete each of the following statements.
   a. Whenever the rotor of the _____ is out of alignment with the rotor of the receiver selsyn, currents are present in the stator windings.
   b. A selsyn receiver rotates continuously if the transmitter rotor is driven at _____ speed.
   c. The rotor units of the transmitter and receiver selsyns must be excited from the same _____.
   d. The use of a mechanical damper on the rotor of selsyn receivers minimizes any tendency of the receiver to _____.

7. List four applications for a split-phase induction motor.
   a. _____
   b. _____
   c. _____
   d. _____

8. Name the basic parts of a split-phase induction motor. _____

   _____

   _____

   _____

   _____

   _____

9. Explain how the direction of rotation of a split-phase induction motor is reversed.

   _____

   _____

   _____

   _____

10. What happens if the centrifugal switch contacts fail to reclose when a split-phase motor is stopped? _____

   _____

   _____

   _____

11. A split-phase induction motor has a dual-voltage rating of 115/230 volts. The motor has two running windings, each of which is rated at 115 volts. The motor also has two starting windings, each of which is rated at 115 volts. Draw a schematic connection diagram of this split-phase induction motor connected for 230 volts.

12. What is the basic difference between a split-phase induction motor and a capacitor-start, induction-run motor? _____

_____

_____

_____

13. If the centrifugal switch fails to open as a split-phase motor accelerates to the rated speed, what happens to the starting winding? _____

_____

_____

_____

14. If the centrifugal switch on a capacitor-start, induction-run motor fails to open as the motor accelerates to the rated speed, what may happen in the starting winding circuit?

_____

_____

_____

_____

15. What is one limitation of a capacitor-start, induction-run motor? _____

_____

_____

_____

16. What is the basic difference between a capacitor-start, induction-run motor and a capacitor-start, capacitor-run motor? _____

_____

_____

_____

17. List three types of capacitor-start, capacitor-run motors.

a. _____

b. _____

c. _____

18. Insert the correct word or phrase to complete each of the following statements.

a. The capacitor in series with the starting winding of a capacitor-start, induction-run motor improves the _____ torque of the motor.

b. A split-phase induction motor has good speed regulation, but _____ starting torque characteristics.

c. A capacitor-start, capacitor-run motor has practically _____ power factor when operating at full load.

d. A capacitor-start, capacitor-run motor has _____ starting torque and _____ speed regulation.

e. A capacitor-start, induction-run motor has _____ speed regulation.

19. Insert the correct word or phrase to complete each of the following statements.
   a. The capacitor used with a capacitor-start, induction-run motor is used only for the purpose of improving the _____ of the motor.
   b. The capacitors used with a capacitor-start, capacitor-run motor are used to improve
      _____.
   c. A motor of 1 hp or less that is manually started and is within sight of the starter location, if the distance is no greater than 50 ft, is considered protected from overload by the _____.
   d. A motor of 1 hp or less that is manually operated but is more than 50 ft from the starter location shall have a _____.
   e. Where separate overload devices are required for motors, they shall not be set at more than _____ % of the motor nameplate, full-load current rating for motors marked to have a temperature rise not over _____ for motors with a marked service factor of 1.15, and at not more than _____ % for all other types of motors.

20. Where is a repulsion motor used?

   _____

   _____

21. What are the two types of repulsion-start, induction-run motors?

   _____

   _____

22. Where are repulsion-start, induction-run motors used?

   _____

   _____

23. A 3 hp, 230-volt, 17-ampere, single-phase, repulsion-start, induction-run motor is connected directly across rated line voltage.
   a. Determine the overcurrent protection for the branch circuit feeding this motor.

   _____

   _____

   _____

b. Determine the running overload protection to use with this motor. _____

_____

_____

_____

24. What size of copper wire (type THHN) is used for the branch circuit feeding the motor in question 23? _____

_____

25. What is a repulsion-induction motor? _____

_____

_____

_____

26. What is one advantage to the use of the repulsion-induction motor as compared with the repulsion-start, induction-run motor? _____

_____

_____

_____

27. Insert the correct word or phrase to complete each of the following statements.
   a. A repulsion motor has good _____ but poor _____.
   b. The speed of a repulsion motor can be controlled by changing the _____.
   c. Both the brush-riding and the brush-lifting types of repulsion-start, induction- after-they-have-run motors operate as _____ after they have accelerated to rated speed.
   d. The repulsion-induction motor has good _____ and relatively good

   _____.

28. Explain how the direction of rotation is changed on any one of the three types of single-phase repulsion motors. _____

_____

_____

_____

29. What is a universal motor?

_____

_____

_____

30. Draw a schematic diagram of a conductively compensated series motor.

31. Draw a schematic diagram of an inductively compensated series motor.

32. In what way is an inductively compensated series motor different from a conductively compensated series motor? _____

_____

_____

_____

33. Explain how the direction of rotation is reversed for a conductively compensated series motor. _____

_____

_____

_____

34. What is the purpose of a compensating winding in an AC series motor?

_____

_____

_____

35. A universal motor can be operated on_____
   a. AC power only.
   b. DC power only.
   c. AC or DC power.

36. A conductively compensated series motor can be operated on _____
   a. AC power only.
   b. DC power only.
   c. AC or DC power.

37. An inductively compensated series motor can be operated on either AC or DC power. _____
   a. true
   b. false

38. A large 25 hp, DC series motor will not operate satisfactorily on an AC power source. _____
   a. true
   b. false

GLOSSARY

across-the-line (ATL)   Method of motor starting that connects the motor directly to the supply line, on starting or running; also called *full voltage control*.

alternating current (AC)   A current that alternates regularly in direction. Refers to a periodic current with successive half-waves of the same shape and area.

alternator   A machine used to generate alternating current by rotating conductors through a magnetic field; an AC generator.

alternator periodic-time relationship   The phase voltages of two generators running at different speeds.

alternators paralleled   Alternators are connected in parallel whenever the power demand of the load circuit is greater than the power output of a single alternator.

ambient temperature   The temperature surrounding a device.

amortisseur winding   Consists of copper bars embedded in the cores of the poles of a synchronous motor. The copper bars of this special type of squirrel-cage winding are welded to end rings on each side of the rotor; used for starting only.

armature   A cylindrical, laminated iron structure mounted on a drive shaft; contains the armature winding.

armature winding   Wiring embedded in slots on the surface of the armature; voltage is induced in this winding on a generator.

automatic compensators   Motor starters that have provisions for connecting three-phase motors automatically across 50%, 65%, 80%, and 100% of the rated line voltage for starting, in that order, after preset timing.

autotransformer   A transformer in which a part of the winding is common to both the primary and secondary circuits.

auxiliary contacts   Contacts of a switching device in addition to the main circuit contacts; auxiliary contacts operate with the movement of the main contacts; electrical interlocks.

blowout coil   Electromagnetic coil used in contactors and motor starters to deflect an arc when a circuit is interrupted.

branch circuit   That portion of a wiring system that extends beyond the final overcurrent device protecting the circuit.

brushless excitation   The commutator of a conventional direct-connected exciter of a synchronous motor is replaced with a three-phase, bridge-type, solid-state rectifier.

brushless exciter   Solid-state voltage control on an alternator, providing DC necessary for the generation of AC.

bus   A conducting bar of different current capacities, usually made of copper or aluminum.

busway   A system of enclosed power transmission that is current and voltage rated.

capacitor   A device made with two conductive plates separated by an insulator or dielectric.

centrifugal switch   On single-phase motors, when the rotor is at normal speed, centrifugal force set up in the switch mechanism causes the collar to move and allows switch contacts to open, removing starting winding.

circuit breaker   A device designed to open and close a circuit by nonautomatic means and to open the circuit automatically on a predetermined overcurrent without injury to itself when properly applied within its rating.

cogenerating plants   Diesel-powered electric generating sets designed to recapture and use the waste heat from both their exhaust and cooling systems.

commutator   Consists of a series of copper segments insulated from one another and the mounting shaft; used on DC motors and generators.

compensator transformer   A tapped autotransformer used for starting induction motors.

conductor   A device or material that permits current to flow through it easily.

conduit plan   A diagram of all external wiring between isolated panels and electrical equipment.

contactor   An electromagnetic device that repeatedly establishes or interrupts an electric power circuit.

controller   A device or group of devices that governs in a predetermined manner the delivery of electric power to apparatus connected to it.

counter-emf (CEMF)   An induced voltage developed in a DC motor while rotating. The direction of the induced voltage is opposite to that of the applied voltage.

cumulative compound-wound generator or motor   A series winding is connected to aid the shunt winding.

current   The rate of flow of electrons measured in amperes.

current flow   The flow of electrons.

DC exciter bus   A bus from which other alternators receive their excitation power.

definite time   A predetermined time lapse.

delta connection A circuit formed by connecting three electrical devices in series to form a closed loop; used in three-phase connections.

diode A two-element device that permits current to flow through it in only one direction.

direct current (DC) Current that does not reverse its direction of flow; a continuous nonvarying current in one direction.

disconnecting switch A switch intended to open a circuit only after the load has been thrown off by some other means; not intended to be opened under load.

drum switch A manually operated switch with electrical connecting parts in the form of fingers held by spring pressure against contact segments or surfaces on the periphery of a rotating cylinder or sector.

dual-voltage motors Motors designed to operate on two different voltage ratings.

duty cycle The period of time in which a motor can safely operate under a load. *Continuous* means that the motor can operate fully loaded 24 hours a day.

dynamic braking Using a DC motor as a generator, taking it off the supply line and applying an energy dissipating resistor to the armature. Dynamic braking for an AC motor is accomplished by disconnecting the motor from the line and connecting DC power to the stator windings.

eddy current Current induced into the core of a magnetic device. Causes part of the iron core losses, in the form of heat.

efficiency The efficiency of all machinery is the ratio of the output to the input—that is, output/input = efficiency.

electric controller A device or group of devices that governs, in some predetermined manner, the electric power delivered to the apparatus to which it is connected.

elementary diagram (ladder diagram, schematic diagram, line diagram) Represents the electrical control circuit in the simplest manner. All control devices and connections are shown as symbols located between vertical lines that represent the source of control power.

emergency generator system A generating set that functions as a power source in a health care facility, such as a hospital; a standby power system. In addition to lighting, the loads supplied are essential to life and safety.

engine-driven generating sets Generators with prime movers of diesel, gasoline, or natural gas engines, and the like.

exciter A DC generator that supplies the magnetic field for an alternator.

feeder The circuit conductor between the service equipment or the switchboard of an isolated plant and the branch-circuit overcurrent device.

field discharge switch Used in the excitation circuit of an alternator. Controls (through a resistor) the high inductive voltage created in the field coils by the collapsing magnetic field.

flux   Magnetic field; lines of force around a magnet.

frequency   Cycles per second or hertz.

fuse   An overcurrent protective device with a circuit-opening fusible part that is heated and severed by the passage of overcurrent through it.

gear motor   A self-contained drive made up of a ball-bearing motor and a speed-reducing gear box.

grounded   Connected to earth or another conducting body that serves in place of earth.

growler   An instrument consisting of an electromagnetic yoke and winding excited from an AC source; used to locate short-circuited motor coils.

hertz   The measurement of the number of cycles of an alternating current or voltage completed in one second.

hysteresis   Part of iron core losses.

identified conductor (neutral)   A grounded conductor in an electrical system, identified with the code color white.

induced current   Current produced in a conductor by the cutting action of a magnetic field.

induced voltage   Voltage created in a conductor when the conductor interacts with a magnetic field.

induction   Induced voltage is always in such a direction as to oppose the force producing it.

insulator   Material with a very high resistance used to electrically isolate two conductive surfaces.

I/O section (input/output section)   This section of a programmable controller interfaces the PC to the electrical signals in the field. The input takes the appropriate field indication and converts it to a signal recognizable by the processor. The output takes a signal sent by the microprocessor and converts it to the proper signal for the field devices.

isolating transformer   A transformer in which the secondary winding is electrically isolated from the primary winding.

jogging   The quickly repeated closure of a controller circuit to start a motor from rest for the purpose of accomplishing small movements of the driven machine.

legally required standby generating systems   Systems required by municipal, state, federal, or other codes or government agencies having jurisdiction.

Lenz's law   A voltage is induced in a coil whenever the coil circuit is opened or closed.

maintaining contact   A small contact in the control circuit used to keep a coil energized, usually actuated by the same coil; also known as a holding contact or an auxiliary contact.

mechanical interlock   A mechanical interlocking device assembled at the factory between forward and reverse motor starters and multispeed starters; it locks out one starter at the beginning of the stroke of either starter to prevent short circuits and burnouts by the accidental closure of both starters simultaneously.

megohmmeter (megger)   An electrical instrument used to measure insulation resistance.

megohms   A unit of resistance equal to 1,000,000 ohms.

motor circuit switch (externally operated disconnect switch, exo)   Motor branch-circuit switch rated in horsepower. Usually contains motor-starting protection; safety switch.

motor controller   A device used to control the operation of a motor.

motorizing   A generator armature rotates as a motor.

motor starter   A device used to start and/or regulate the current to a motor during the starting period. It may be used to make or break the circuit and/or limit the starting current. It is equipped with overload protection devices, such as a contactor with overload relays.

multimeter   Electrical instrument designed to measure two or more electrical quantities.

NEC®   *National Electrical Code*®.

nonsalient rotor   A rotor that has a smooth cylindrical surface. The field poles (usually two or four) do not protrude above this smooth surface.

normal field excitation   The value of DC field excitation required to achieve unity power factor in a synchronous motor.

normally open, normally closed   When applied to a magnetically operated switching device (such as a contactor or relay, or to the contacts of these devices), these terms signify the position taken when the operating magnet is de-energized, and with no external forces applied. The terms apply only to nonlatching types of devices.

ohmmeter   An instrument used to measure resistance.

oil (immersed) switch   Contacts of a switch that operate in an oil bath tank. Switch is used on high voltages to connect or disconnect a load. Also known as an oil circuit breaker.

overload   Operation of equipment in excess of normal, full-load rating, or of a conductor in excess of rated ampacity which, when it persists for a sufficient length of time, would cause damage or dangerous overheating.

overload protection (running protection)   Overload protection is the result of a device that operates on excessive current, but not necessarily on a short circuit, to cause the interruption of current flow to the device governed.

parallel circuit   A circuit that has more than one path for current flow.

permeability   The ease with which a material conducts magnetic lines of force.

plugging   Braking a motor by reversing the line voltage or phase sequence; motor develops a retarding force; a quick stop.

plugging relay   A device attached to a motor shaft to accomplish plugging switches reversing starter to establish countertorque that brings the motor to a quick standstill before it begins to rotate in the reverse direction.

polarity   Characteristic (negative or positive) of a charge. The characteristic of a device that exhibits opposite quantities, such as positive and negative, within itself.

pole   The north or south magnetic end of a magnet; a terminal of a switch; one set of contacts for one circuit of main power.

polyphase   An electrical system with the proper combination of two or more single-phase systems.

power factor   The ratio of true power to apparent power. A power factor of 100% is the best electrical system.

preventive maintenance   Periodic inspections to prevent serious damage to machinery by locating potential trouble areas; preventing breakdowns rather than repairing them.

processor   The microprocessor section of a programmable controller. It is the section of a PC that holds the programs, receives information, makes a decision, and delivers an output signal to some external electrical device.

programmable controller (PC)   A microprocessor-based system used to control electrical operations. It is a control system controlled by software that can easily be altered to provide flexible control schemes.

programmer   A device—either hand held or personal computer, or special monitor and keyboard—that allows a person to enter desired control programs to the microprocessor section of a PC.

pulse-width modulation (PWM)   A process that controls the width of a pulse delivered to an AC motor. By modulating the width of several pulses and controlling the amplitude, a waveform that approximates a sinewave is produced at adjustable frequencies.

pushbutton   A master switch; manually operated plunger or button for an electrical actuating device; assembled into pushbutton stations.

raceway   A channel or conduit designed expressly for holding wires, cables, or busbars.

rating   The rating of a switch or circuit breaker includes (1) the maximum current and voltage of the circuit on which it is intended to operate, (2) the normal frequency of the current, and (3) the interrupting tolerance of the device.

rectifier   A device that converts AC into DC.

regulation   Voltage at the terminals of a generator or transformer for different values of the load current; usually expressed as a percentage.

relay   Used in control circuits; operated by a change in one electrical circuit to control a device in the same circuit or another circuit.

remote control Controls the function initiation or change of an electrical device from some remote place or location.

residual flux A small amount of magnetic field.

resistance starter (primary resistance starter) A controller to start a motor at a reduced voltage with resistors in the line on start.

rheostat A resistor that can be adjusted to vary its resistance without opening the circuit in which it may be connected.

r/min (RPM) Speed in revolutions per minute.

rotor The revolving part of an AC motor or alternator.

salient field rotor Found on three-phase alternators and synchronous motors; field poles protrude from the rotor support structure. The structure is of steel construction and commonly consists of a hub, spokes, and a rim. This support structure is called a spider.

SCADA Acronym for supervisory control and data acquisition.

selsyn Abbreviation of the term *self-synchronous*. Selsyn units are special AC motors used primarily in applications requiring remote control. These units are also referred to as *synchros*.

semiconductors Materials that are neither good conductors nor good insulators. Certain combinations of these materials allow current to flow in one direction but not in the opposite direction.

separate control The coil voltages of a relay, contactor, or motor starter are separate or different from those at the switch contacts.

separately excited field The electrical power required by the field circuit of a DC generator may be supplied from a separate or outside DC supply.

series field In a DC motor, has comparatively few turns of wire of a size that permits it to carry the full-load current of the motor.

series winding Generator winding connected in series with the armature and load; carries full load.

service factor An allowable motor overload; the amount of allowable overload is indicated by a multiplier which, when applied to a normal horsepower rating, indicates the permissible loading.

short and ground A flexible cable with clamps on both ends used to ground and short high lines to prevent electrical shock to workers.

shunt To connect in parallel; to divert or be diverted by a shunt.

silicon-controlled rectifier (SCR) A four-layer semiconductor device that is a rectifier. It must be triggered by a pulse applied to the gate before it will conduct electricity.

single phase A term characterizing a circuit energized by a single alternating EMF. Such a circuit is usually supplied through two wires.

slip　In an induction motor, slip is the difference between the synchronous speed and the rotor speed, usually expressed as a percentage.

slip rings　Copper or brass rings mounted on, and insulated from, the shaft of an alternator or wound-rotor induction motor; used to complete connections between a stationary circuit and a revolving circuit.

solenoid　An electromagnet used to cause mechanical movement of an armature, such as a solenoid valve.

solid state　As used in electrical-electronic circuits, refers to the use of solid materials as opposed to gases, as in an electron tube. It usually refers to equipment using semiconductors.

speed control　Refers to changes in motor speed produced intentionally by the use of auxiliary control, such as a field rheostat or automatic equipment.

speed regulation　Refers to the changes in speed produced by changes within the motor due to a load applied to the shaft.

split phase　A single-phase induction motor with auxiliary winding, displaced in magnetic position from, and connected parallel to, the main winding.

standby power-generating system　Alternate power system for applications such as heating, refrigeration, data processing, or communications systems where interruption of normal power would cause human discomfort or damage to a product in manufacture.

starting current　The surge of amperes of a motor upon starting.

starting protection　Overcurrent protection is provided to protect the motor installation from potential damage due to short circuits, defective wiring, or faults in the motor controller or the motor windings. The starting protection may consist of a motor disconnect switch containing fuses.

stator　The stationary part of a motor or alternator; the part of the machine that is secured to the frame.

synchronous alternators　Frequencies, voltages, and instantaneous AC polarities must be equal when paralleling machines.

synchronous capacitor　A synchronous motor operating only to correct the power factor and not driving any mechanical load.

synchronous motor　A three-phase motor (AC) that operates at a constant speed from a no-load condition to full load; has a revolving field that is separately excited from a DC source; similar in construction to a three-phase AC alternator.

synchronous speed　The speed at which the electromagnetic field revolves around the stator of an induction motor. The synchronous speed is determined by the frequency (hertz) of the supply voltage and the number of poles on the motor stator.

synchroscope　An electrical instrument for synchronizing two alternators.

tachometer　An instrument used to check the speed of a motor or machine.

three phase   A term applied to three alternating currents or voltages of the same frequency, type of wave, and amplitude. The currents and/or voltages are one-third of a cycle (120 electrical time degrees) apart.

three-phase system   Electrical energy originates from an alternator that has three main windings placed 120° apart. Three wires are used to transmit the energy.

thyristor   An electronic component that has only two states of operation—on or off.

torque   The rotating force of a motor shaft produced by the interaction of the magnetic fields of the armature and the field poles.

transfer switches   Switches to transfer, or reconnect, the load from a preferred or normal electric power supply to the emergency power supply.

transformer   An electromagnetic device that converts voltages for use in power transmission and operation of control devices.

transformer bank   When two or three transformers are used to step down or step up voltage on a three-phase system.

transformer secondary winding   The coil that discharges the energy at a transformed or changed voltage, up or down.

wheatstone bridge   Circuit configuration used to measure electrical qualities such as resistance.

wiring diagram   Locates the wiring on a control panel in relationship to the actual location of the equipment and terminals; made up of a method of lines and symbols on paper.

wound-rotor induction motor   An AC motor consisting of a stator core with a three-phase winding, a wound rotor with slip rings, brushes and brush holders, and two end shields to house the bearings that support the rotor shaft.

wye connection (star)   A connection of three components made in such a manner that one end of each component is connected. This connection generally connects devices to a three-phase power system.

INDEX